12th Europhysical Conference on Defects in Insulating Materials

(EURODIM 2014)

IOP Conference Series: Materials Science and Engineering
Volume 80

Canterbury, UK
13-19 July 2014

ISBN: 978-1-5108-0294-0
ISSN: 1757-8981

Printed from e-media with permission by:

Curran Associates, Inc.
57 Morehouse Lane
Red Hook, NY 12571

Some format issues inherent in the e-media version may also appear in this print version.

Printed by Curran Associates, Inc. (2015)

For permission requests, please contact the Institute of Physics
at the address below.

Institute of Physics
Dirac House, Temple Back
Bristol BS1 6BE UK

Phone: 44 1 17 929 7481
Fax: 44 1 17 920 0979

techtracking@iop.org

Additional copies of this publication are available from:

Curran Associates, Inc.
57 Morehouse Lane
Red Hook, NY 12571 USA
Phone: 845-758-0400
Fax: 845-758-2634
Email: curran@proceedings.com
Web: www.proceedings.com

TABLE OF CONTENTS

IOP Conf. Series: Materials Science and Engineering **80** (2015) 011001 doi:10.1088/1757-899X/80/1/011001

PREFACE

The 12th Europhysical Conference on Defects in Insulating Materials (EURODIM14) was held at the University of Kent, UK, from 13-18 July 2014. It was attended by about 120 delegates from around the world, and featured 56 oral presentations and 77 posters.

EURODIM14 followed other conferences in the series, held in Pecs (2010), Milan (2006) and Wroclaw (2002), as well as the related ICDIM conferences held in Santa Fe (2012), Aracaju (2008) and Riga (2004). These conferences all have the aim of bringing together scientists to discuss the chemistry and physics of defects in solids, and their role in determining material properties.

We would like to thank the International Advisory Committee for suggesting invited speakers, and the Local and Programme Committee for their hard work in planning and running the conference. Finally we would like to thank the authors and referees for their contributions to the proceedings.

M L Alfredsson (Conference Chair)
A V Chadwick
R A Jackson
E E McCabe

IOP Conf. Series: Materials Science and Engineering **80** (2015) 011002 doi:10.1088/1757-899X/80/1/011002

List of Delegates

Dr Abdirash Akilbekov
6/5, Abylaikhan av., 52 apt.
Astana, 010008
Kazakhstan

Dr Antonino Alessi
University of Palermo
Department of Physics and Chemistry,
Via Archirafi 36
Palermo, I-90123
Italy

Dr Maria Alfredsson
University of Kent
School of Physical Sciences
Canterbury, CT2 7NH
Kent, UK

Mrs Meaad Al-hadidi
Newcastle University
286 Stanhope Street
Newcastle upon Tyne, NE4 5JU
UK

Mr Zhunusbekov Amangeldy
L.N. Gumilyov Eurasian National University
Mirzoyan st., 2,
Astana, 010008
Kazakhstan

Mr Adriano Andrade
Federal University of Sergipe
Rua F, N°102, Eduardo Gomes
São Cristovão, 49100-000
Brazil

Dr Donna Arnold
University of Kent
School of Physical Sciences
Canterbury, CT2 7NH
Kent, UK

MSc Andreas Artac
University of Vienna
Beckmanngasse 40
Wien, 1150
Austria

Dr Dmitry Azamat
Institute of Physics
Academy of Sciences of the Czech Republic
Na Slovance 2
Prague, 182 21
Czech Republic

Dr Marie-France Barthe
CNRS CEMHTI Site Cyclotron
3 A rue de la Férollerie
Orléan, 45071
France

Dr Marie-Geneviève Blanchin
ILM, Université Claude-Bernard
Université de LYON
Campus de la Doua
Villeurbanne, 69622
FRANCE

Mr Samuel Bradley
University College London
Gower Street
London, WC1E 6BT
UK

Prof Frank Bridges
University of California,
144 La Canada Way
Santa Cruz, CA
95060, United States

Prof Zdenek Bryknar
Czech Technical University in Prague
Trojanova 13
Prague 2, CZ-12000
Czech Republic

Ph. D. Maksym Buryi
Institute of Physics AS CR, v.v.i (FZU)
Cukrovarnická 10/112
Prague, 162 00
Czech Republic

Mr Nilo Cano-Mamani
UNIFESP
Rua Francisco dos Santos 173
Sao Paulo, 66318
Brazil

Prof Alan V. Chadwick
University of Kent
School of Physical Sciences
Canterbury, CT2 7NH
Kent, UK

Mrs Catherine Comins
24 Maple Avenue
Plantation
Boksburg, 1459
South Africa

Prof Darrell Comins
School of Physics
University of the Witwatersrand
Johannesburg
Wits 2050
South Africa

Mr Matthew Coomber
University of Kent
School of Physical Sciences
Canterbury, CT2 7NH
Kent, UK

Prof Alastair Cormack
Alfred University
NY State College of Ceramics at Alfred University
2 Pine St
Alfred, NY
USA

Dr Gábor Corradi
Wigner Research Centre for Physics
P.O.Box 49
Budapest, H-1525
Hungary

Mr Jonathon Cottom
Dept Physics & Astronomy
University College London
Gower Street
London, WC1E 6BT
UK

Dr Nishith Das
University College London
Christopher Ingold Building
20 Gordon Street
London, WC1H 0AJ
UK

Dr Alma Dauletbekova
6/5, Abylaikhan av., 52 apt.
Astana, 010008
Kazakhstan

Dr James A Dawson
Kyoto University
204 San Flora Shimogamo East
Takanoshimizucho
Sakyo, 606-8102
Kyoto-fu
Japan

Mr Oliver Dicks
Department of Physics & Astronomy
University College London
London, WC1E 6BT
UK

Prof Volkmar Dierolf
Lehigh University
16 Memorial Drive East
Bethlehem, 18015
USA

Dr Andre Düvel
Leibniz Universität
Callinstr. 3-3a
Hannover, 30167
Germany

Mr Al-Moatasem El-Sayed
University College London
39 Crescent Way
London, N12 0RD
UK

Dr Mauro Fasoli
Department of Materials Science
Universtity of Milano Bicocca
via Cozzi 55
Milan, I-20125
Italy

Mr Giordano Frederico
Federal University Sergipe
São Luiz Street, 828
Itabaiana-SE, 49500-000
Brazil

Dr Juan María García Lastra
Technical University of Denmark
Fysikvej. Building 311
Kongens Lyngby, 2800
Denmark

Dr Corneliu Ghica
National Institute of Materials Physics, Romania
Atomistilor 105 bis
Magurele
Ilfov, 077125
Bucharest
Romania

Dr Daniela Ghica
National Institute of Materials Physics
Magurele
Ilfov, 077125
Bucharest
Romania

Dr Luigi Giacomazzi
CNR-IOM Democritos
via Bonomea 265
Trieste, IT-34136
Italy

Mr Manassés Gomes
UFS
Rua Grujim, Rosa Elze, 430, CASA 06
São Cristovão, 49100-000
Sergipe
Brazil

Dr Partha Goswami
Deshbandhu College, Kalkaji
University of Delhi
New Delhi, 110019
Delhi,
India

Prof Dr Siegmund Greulich-Weber
University of Paderborn
Delpstraße 22
Bad Lippspringe, 33175
Germany

Ms Ivett Hajdara
Wigner RC
Konkoly-Thege M. 29-33
Budapest, 1121
Hungary

Mr Matthew Halliday
UCL
75A Denmark Hill
London, SE5 8RS
UK

Mr. Daniel Jackson
University of Kent
School of Physical Sciences
Canterbury, CT2 7NH
Kent, UK

Dr Robert Jackson
Lennard-Jones Laboratories, School of Physical & Geographical Sciences
Keele University
Keele, ST5 5BG
Staffordshire, UK

Dr Lawrence Kappers
University of Connecticut
600 N. 6th St.
Oostburg, 53070
Wisconsin, USA

Dr William Karstens
Saint Michael's College
One Winooski Park
Colchester, 05439
Vermont, USA

Mr Kenneth Kgatwane
University of Limpopo
P.O. Box 76
Baratheo
Rustenburg, 0338
Republic of South Africa

Ms Tinku Khanwalkar
208 N Marshal St
Allentown, 18104
PA, USA

Ms Elena Khudyakova
Tomsk State University of Control Systems and Radioelectronics (TUSUR)
Lenin avenue 40
Tomsk, 634050
Russia

IOP Conf. Series: Materials Science and Engineering **80** (2015) 011002 doi:10.1088/1757-899X/80/1/011002

Prof Nobuhiro Kodama
1-1, Tegata-gakuencho
Akita, 010-8502
Japan

Mr Valdis Korsaks
Institute of Solid State Physics
University of Latvia
8 Kengaraga street
Riga, LV1063
Latvia

Dr Laszlo Kovacs
Wigner Research Centre
Konkoly-Thege út 29-33
Budapest, 1121
Hungary

Ms Anna Kozlova
B-49, Leninsky prospect, 4
Moscow, 119049
Russia

Dr Klaus Krambrock
UFMG
Av. Antônio Carlos, 6627
Pampulha
Belo Horizonte – MG, 31270-901
Brazil

Mr Aleksei Krasnikov
Institute of Physics, University of Tartu
Riia 142
Tartu, 51014
Estonia

Prof Toshio Kurobori
Kanazawa University
Graduate School of Natural Science
and Technology
Kakuma, 920-1192
Japan

Mr Tero Laihinen
University of Turku
Savitehtaankatu 2B 61
Turku, 20540
Finland

Dr Milan Lalic
UFS – Brazil
Rua Sônia Alves Lopes, 2681
Aracaju, 49035-740
Brazil

Ms Raesibe Sylvia Ledwaba
Materials Modelling Centre
School of Physical and Mineral Sciences
University of Limpopo
Private Bag x 1106
Sovenga, 0727
Polokwane
South Africa

Mr Krisztián Lengyel
Wigner Research Centre
Konkoly-Thege Miklós út 29-33
Budapest, 1121
Hungary

Dipl.-Min Christoph Lenz
Institut für Mineralogie und Kristallographie
Universität Wien
Althanstraße 14 (UZA2)
Wien, 1090
Austria

Dr Vladyslav Litichevskyi
Institute for Scintillation Materials NAS of Ukraine
Lenin ave. 60
Kharkiv, 61001
Ukraine

Prof Aleksandr Lushchik
Institute of Physics
University of Tartu
Tartu, 51014
Estonia

Dr Zelia Macedo
Federal Univerisity of Sergipe
R Benjamin Fontes 117 Luzia
Aracaju, 49045110
Sergipe, Brazil

Dr Patrick Martin
CELIA – CNRS
351 cours de la Libération
Talence, 33405
France

Ing. Jiri Martincik
Institute of Physics of the Academy of Sciences of the Czech Republic
Cukrovarnicka 10
Prague, 162 00
Czech Republic

Dr Malili Matshaba
University of Limpopo
P.O. Box 638
Senwabarwana, 0790
Polokwane
South Africa

Dr Emma McCabe
University of Kent
School of Physical Sciences
Canterbury, CT2 7NH
Kent, UK

Dr Yoshimi Mita
Graduate School of Engineering Science
Osaka University
Toyonaka
Osaka, 560-8531
Japan

Mr Juan Mittani
UNIFESP
Alameda Barao de Limeira, 89
Apto 605
Sao Paulo, 01202001
SP, Brazil

Dr Rosa Maria Montereali
ENEA C.R. Frascati
V. E. Fermi, 45
Frascati, 00044
Italy

Prof Miguel Moreno
University of Cantabria
Facultad de Ciencias
Santander, 39005
Spain

Dr Federico Moretti
Institut Lumiére Matiére
Universitè Claude Bernard Lyon 1
20 rue Ada Byron
Villeurbanne, 69622 CEDEX
France

Dr Toshiya Murakami
Wakayama University
Sakaedani 930
Wakayama-city, 640-8510
Japan

Dr Andrey Mysovsky
Irkutsk State Technical University
83 Lermontov Street
Irkutsk, 664074
Irkutsk region
Russia

Mrs Raluca Florentina Negrea
National Institute of Materials Physics Atomistilor
Str. 105 bis
Magurele
Bucharest, 077125
Ilfov, Romania

Prof Phuti Ngoepe
Materials Modelling Centre
University of Limpopo
Private Bag x1106
Sovenga, 0727
Limpopo Province, South Africa

Dr Leona Cristina Nistor
National Institute of Materials Physics Atomistilor
Str. 105 bis
Magurele
Bucharest, 077125
Ilfov, Romania

Dr Vasile Sergiu Nistor
National Institute of Materials Physics Atomistilor
Str. 105 bis
Magurele
Bucharest, 077125
Ilfov, Romania

Prof Alexander Ogurtsov
National Technical University
Frunse Str., 21
Kharkov, 61002
Ukraine

Miss Reeya Oogarah
University of Kent
School of Physical Sciences
Canterbury, CT2 7NH
Kent, UK

IOP Conf. Series: Materials Science and Engineering **80** (2015) 011002 doi:10.1088/1757-899X/80/1/011002

Msc Annika Pille
Institute of Physics
University of Tartu
Riia 142
Tartu, 51014
Estonia

Prof Erasmus Rammutla
University of Limpopo
Physics Department
P/Bag x1106
Sovenga, 0727
Polokwane, South Africa

Miss Marine Riesenmey
University of Kent
School of Physical Sciences
Canterbury, CT2 7NH
Kent, UK

Prof Aline Rougier
ICMB,
CNRS
UPR9048
Pessac
France

Mrs Ana Santos
Cidade Universitária
Prof. José Aloísio de Campos
Av. Marechal Rondon
s/n Jardim Rosa Elze
São Critovão, CEP 49100-000
Brazil

Dr Dean Sayle
University of Kent
School of Physical Sciences
Canterbury, CT2 7NH
Kent, UK

Prof Alex Shluger
Department of Physics & Astronomy
University College London
London, WC1E 6BT
UK

Mr Petr Prusa
Institute of Physics ASCR
Cukrovarnicka 10
Prague, 16200
Czech Republic

Dr Mark Read
School of Chemistry
University of Birmingham
Edgbaston
Birmingham, B15 2TT
UK

Ivo Romet
Institute of Physics
University of Tartu
Riia 142
Tartu, 51014
Estonia

Dr Brian Sales
Oak Ridge National Laboratory
355 Elizabeth Way
Lenoir City, 37772
TN, United States

David Sampaio
UFS
St. Floduardo R. Dorea 66
Aracaju, 49095-130
Brazil

Ms Ashley Shields
UCL
Department of Chemistry
20 Gordon Street
London, WC1H 0AJ
UK

Dr Kurt Sickafus
The University of Tennessee
414F Ferris Engineering Bldg.
1508 Middle Way Drive
Knoxville, 37996-2100
Tennessee, United States

IOP Conf. Series: Materials Science and Engineering **80** (2015) 011002 doi:10.1088/1757-899X/80/1/011002

Prof Maria Silva
Departamento de Engenharia Cerâmica e do Vidro, Campus
Universitário de Santiago,
Universidade de Aveiro
Aveiro, 3810-193
Portugal

Dr Vera Skvortsova
Institute of Solid State Physics
University of Latvia
8 Kengaraga Str.,
Riga, LV-1063
Latvia

Dr Philippe Smet
LumiLab
Ghent University
Krijgslaan 281-S1
Gent, 9000
Belgium

Ms Barbara Smith
University of Vermont
16 Summit Circle
Shelburne, 05482
Vermont, United States

Prof David Y. Smith
University of Vermont & Argonne National Laboratory
16 Summit Circle
Shelburne, 05482
Vermont, United States

Dr Chris Stanek
Los Alamos National Laboratory
MS G755
Los Alamos, 87545
United States

Mr Jimmel Stewart
University of Kent
School of Physical Sciences
Canterbury, CT2 7NH
Kent, UK

Miss Rebecca Tanney
University of Kent
School of Physical Sciences
Canterbury, CT2 7NH
Kent, UK

Prof François Treussart
Ecole Normale Supérieure de Cachan
61 avenue du président Wilson
Cachan
France

Dr Olexandr Tychko
Taras Shevchenko National University of Kyiv
64/13 Volodymyrska Street
Kyiv, 01601
Ukraine

Dr Mario Valerio
Federal University of Sergipe
R Benjamin Fontes 117 Luzia
Aracaju, 49045110
Sergipe, Brazil

Professor Anna Vedda
Department of Materials Science
Universitry of Milano-Bicocca
via Cozzi 55
Milan, 20125
Italy

Miss Laura Vera-Stimpson
University of Kent
School of Physical Sciences
Canterbury, CT2 7NH
Kent, UK

Miss Irene Villa
Università degli Studi di Milano Bicocca
Via Roberto Cozzi 55
Milano, 20125
Italy

Mrs Rosalba Villarreal
142 Arbour Crest Circle NW
Calgary AB, T3G 4H3
Canada

Dr J Eduardo Villarreal-Barajas
University of Calgary
1331 29 Street NW
Calgary AB, T2N 4N2
Canada

IOP Conf. Series: Materials Science and Engineering **80** (2015) 011002 doi:10.1088/1757-899X/80/1/011002

Dr Henk Vrielinck
Ghent University
Dept. Solid State Sciences
Krijgslaan 281, S1
Gent, B-9000
Belgium

Mr Scott Walker
Keele University
Lennard-Jones Laboratories
School of Physical & Geographical Sciences
Keele, ST5 5BG
Staffordshire, UK

Mrs. Renate Watanabe
University of São Paulo
Institute of Physics
Rua Alvarenga 715
Sao Paulo, 66318
Brazil

Mr. Shigueo Watanabe
University of São Paulo
Institute of Physics
Rua Alvarenga 715
Sao Paulo, 66318
Brazil

Prof Graeme Watson
Trinity College
The University of Dublin
School of Chemistry
Dublin, 2
Ireland

Dr Annette Weber
University of Paderborn
Delpstraße 22
Bad Lippspringe, 33175
Germany

Dr Karl Whittle
University of Sheffield
Dept of Materials Science and Engineering
Mappin Street
Sheffield, S1 3JD
South Yorkshire, UK

Mr Michael Woods
University of Kent
School of Physical Sciences
Canterbury, CT2 7NH
Kent, UK

IOP Conf. Series: Materials Science and Engineering **80** (2015) 011003 doi:10.1088/1757-899X/80/1/011003

Peer review statement

All papers published in this volume of *IOP Conference Series: Materials Science and Engineering* have been peer reviewed through processes administered by the proceedings Editors. Reviews were conducted by expert referees to the professional and scientific standards expected of a proceedings journal published by IOP Publishing.

IOP Conf. Series: Materials Science and Engineering **80** (2015) 012001 doi:10.1088/1757-899X/80/1/012001

Non-destructive readout of 2D and 3D dose distributions using a disk-type radiophotoluminescent glass plate

T Kurobori[1], Y Maruyama[1, 2], Y Miyamoto[3], T Sasaki[1] and H Nanto[4]

[1] Graduate School of Natural Science and Technology, Kanazawa University, Kakuma, Kanazawa 920-1192, Japan
[2] Pulstec Industrial Corporation, 7000-35 Nakagawa, Hosoe, Kita-ku, Hamamatsu 431-1304, Japan
[3] Oarai Research Center, Chiyoda Technol Corporation, 3681 Narita-cho, Oaraimachi, Ibaraki 311-1313, Japan
[4] Adv. Mater. Sci. R&D Center, Kanazawa Institute of Technology, 3-1 Yatsukaho, Hakusan 924-0838, Japan

E-mail: kurobori@staff.kanazawa-u.ac.jp

Abstract. Novel disk-type X-ray two- and three-dimensional (2D, 3D) dose distributions have been developed using atomic-scale defects as minimum luminescent units, such as radiation-induced silver (Ag)-related species in a Ag-activated phosphate glass. This luminescent detector is based on the radiophotoluminescence (RPL) phenomenon. Accurate accumulated dose distributions with a high spatial resolution on the order of microns over large areas, a wide dynamic range covering three orders of magnitude and a non-destructive readout were successfully demonstrated for the first time by using a disk-type glass plate with a 100-mm diameter and a 1-mm thickness. In addition, the combination of a confocal optical detection system with a transparent glass detector enables 3D reconstruction by piling up each dose image at different depths within the material.

1. Introduction

The passive dosimeters based on the radiophotoluminescence (RPL) [1-3], optically stimulated luminescence (OSL) [4, 5], photoluminescence (PL) [6-8] and thermoluminescence (TL) [9] phenomena have been widely used for personal, environmental and clinical dosimetry. Although the four types of passive luminescent dosimeters each have advantages and disadvantages [9-11], few reports for 2D and 3D dose distributions on large areas have been published. A dosimeter made of silver (Ag)-activated phosphate glass, which is known as the most common RPL material, has been recognized as possessing several desirable characteristics, such as non-destructive readout capabilities, long-term stability against fading, wide dynamic range and uniformity/batch homogeneity [12].

Recently, a novel disk-type 2D imaging detector with the aforementioned superior characteristics of the RPL dosimeter (referred to as "the glass batch") was proposed and demonstrated [13] for the first time. The 2D dose distribution written in the form of atomic-scale luminescent defects by X-ray irradiation was rapidly reconstructed with a high spatial resolution of 1 μm and a sensitivity of 1 mGy, when the disk-type detector with a diameter of 80 mm was rotated at a rate of 400 rpm (6.7 Hz) to read out the accumulated dose information.

In addition, a comparative study of 2D dose images from Ag-activated phosphate glass using the Ag^{2+} colour centres (CCs) and LiF thin films using F-aggregate CCs based on the RPL and PL phenomena was performed, respectively, and their performances were also evaluated [14]. The 2D dose images with a wide dynamic range covering eight orders of magnitude were successfully demonstrated by combining a Ag-doped glass detector with a LiF thin film detector deposited on a glass substrate. These disk-type detectors were rotated at a rate of 2400 rpm (40 Hz). A detailed description of the optical and dosimetric properties of Ag-related defects in Ag-activated phosphate glass has been published elsewhere [15, 16].

In this paper, first, the optical properties of X-ray-irradiated Ag-activated phosphate glass were studied under absorbed doses from 0.1 to 10 Gy. Specifically, the effects of the absorption rate at an exciting wavelength of 371 nm and the re-absorption rate at an emitting wavelength of 630 nm (i.e., orange RPL) were evaluated from the optical absorption spectra. Second, an accurate 2D dose distribution with a high spatial resolution, wide dynamic range and non-destructive readout was successfully demonstrated using a disk-type Ag-activated phosphate glass plate. Finally, the use of a confocal detection system and a transparent glass detector allows one to reconstruct a 3D dose distribution by combining each image at different depths from the vicinity of the surface to ≈400 μm.

2. Experimental procedure

A commercially available Ag-activated phosphate glass was used as 2D and 3D RPL detectors. Although the weight composition of the used detector was the same as that of the FD-7 (Asahi Techno Glass), i.e., 31.55% P, 51.16% O, 6.12% Al, 11.00% Na and 0.17% Ag, the size and shape were different; the detector was a disk-type plate with a diameter of 100 mm and a thickness of 1 mm.

In this work, the samples were coloured by irradiation from an X-ray unit (8.05 keV) with a copper target operating at 30 kV and 20 mA. The absorbed doses on the samples ranged from 0.1 to 10 Gy. The dose rate from the X-ray unit was calibrated with a commercially available Ag-activated phosphate glass dosimeter. The irradiation dose measurements were performed with a technique based on a combination of a pulsed UV laser excitation and a photon counting system at the Oarai Research Centre of the Chiyoda Technol Corporation, resulting in a dose rate of 117 mGy/min at a distance of 320 mm from the X-ray tube.

Optical absorption (OA) was obtained at room temperature (RT) using a Hitachi U-3900H UV-VIS. The RPL spectra were obtained by a Hitachi F-2500 fluorescence spectrophotometer. For all of the optical measurements, the samples were cut to a suitable size from the original glass dosimeter plate in approximately $10 \times 7 \times 1\text{-mm}^3$ rectangular plates. The radiative lifetime measurements of the blue and orange RPL were performed using a fluorescence lifetime spectrometer (Quantaurus-Tau C11367-14, Hamamatsu Photonics) at RT. Light that was 340 nm from an LED source was used as an excitation beam. The total time resolution was less than 1.0 ns.

3. Results and discussion

Figures 1(a) and (b) show the OA and RPL spectra in Ag-activated phosphate glass with a thickness of 1 mm, respectively, before and after X-ray irradiation under absorbed doses from 0.1 to 10 Gy. Each Ag-activated glass plate was heated to 100 °C for 15 min to suppress the "build-up" kinetics after X-ray irradiation. A set of absorption spectra were attributed to the superposition of a number of individual absorption bands corresponding to the atomic Ag atoms, charged Ag-aggregate CCs and neutral Ag molecular clusters and phosphorous- and oxygen-related species in the range from 250 to 700 nm, as already reported [15, 16]. The primary band peaking at 310 nm emits the orange RPL at 560 nm, which is attributed to the Ag^{2+} centres. Another band peaking at 340 nm emits the blue RPL at 450 nm, which is attributed to the Ag^{0} centres. In this work, the blue and orange RPL were simultaneously emitted by an excitation wavelength at 371 nm because of the superposition of the blue and orange absorption bands. Note that both peak bands for the blue and orange RPL were shifted to the longer wavelengths from 450 to 470 nm and from 560 to 630 nm, respectively, by a UV laser excitation. In addition, the dosimetric relationship between the response of the blue and orange RPL intensity and X-ray absorbed doses from 0.1 to 10 Gy exhibited a good dose linearity response. From Fig. 1(a), we estimated the values of the absorption coefficient and re-absorption rate when light at 630 nm propagates a 500-μm length in the X-ray-irradiated glass under an absorbed dose of 1 Gy. The values are 1.15×10^{-5} (μm^{-1}) and 0.57 %, respectively. For an excitation wavelength at 371 nm, the corresponding values are 1.28×10^{-4} (μm^{-1}) and 6.2 %, respectively. To acquire accurate 3D dose distributions, one must take these values into consideration.

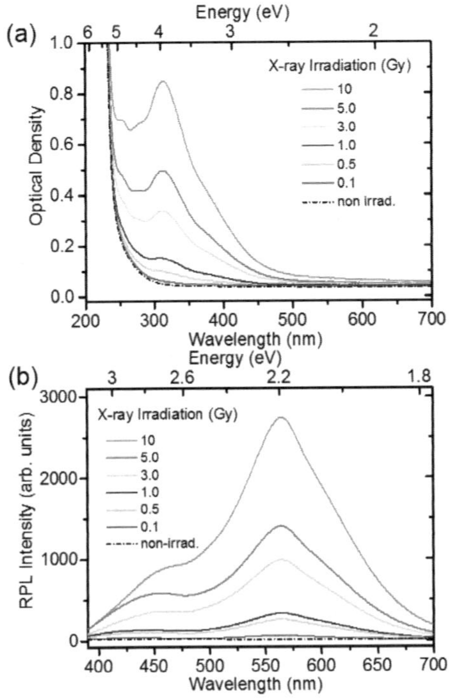

Figure 1. (a) OA spectra of Ag-activated phosphate glass with a thickness of 1 mm before and after X-ray irradiation with absorbed doses from 0.1 to 10 Gy. (b) Corresponding RPL spectra excited at 340 nm before and after X-ray irradiation at RT.

A schematic view of the experimental arrangement employed to measure the 2D and 3D dose distributions is shown in Fig. 2. Collimated light at 371 nm from a cw laser diode (LuxX 371-70, Omicron GmbH) with a maximum power of 70 mW was used as an excitation source. For reading, a low-level power between 10 and 20 mW was used for stimulating the orange RPL at 630 nm. The laser beam was expanded and reflected with a dichroic beam splitter (B/S) (#86-330, Edmund Optics) and focused in the vicinity of the surface for the disk-type detector by a Nikon LU Plan Fluor objective lens (100×, NA=0.90) while reading out the image. The objective lens was simultaneously used for both laser excitation and emission collection. Each detector was attached to a spindle on a linear translation stage, which was rotated at a rate of 2400 rpm and controlled to translate the laser beam spot from the outer disk to the inner disk in the radial direction with a pitch of 20 μm. Additional optical filters, including band-pass (BP)(#86-736, 375±5 nm, Edmund Optics), long-pass (LP) (#62-981, Edmund Optics), BP for the orange RPL (#84-797, 600±25 nm, Edmund Optics) and BP for the blue RPL (#86-962, 475±25 nm, Edmund Optics), were inserted to reject the reflected emission and residual stimulating laser light, respectively. Only the signal was detected by the photomultiplier tube (PMT) (#H10720-20, Hamamatsu Photonics), which then digitized the signal using a 500-kSamples/s, 16-bit analogue-to-digital (A/D) converter. The dose distribution was reconstructed by a personal computer. The total readout and reconstructed times were one to five min, depending on the track pitch (10 to 100 μm) and the number of tracks. The glass detector was irradiated with absorbed doses from 1 to 3 Gy. After X-ray irradiation, the RPL detector was preheated at 100 °C for 15 min before measuring.

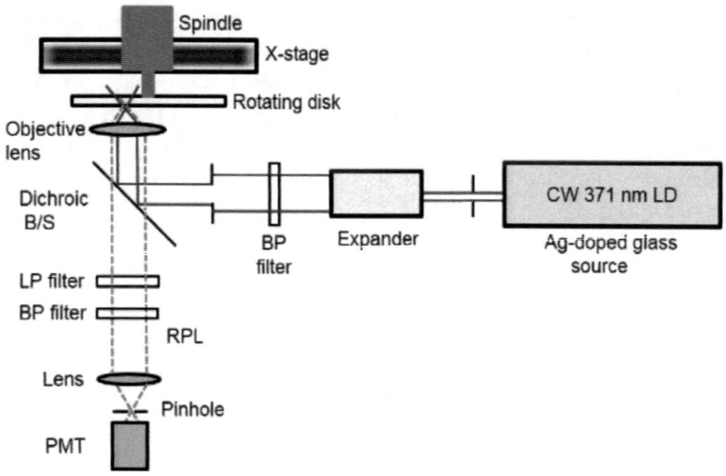

Figure 2. Schematic diagram of a disk-type RPL glass detector and a confocal optical detection system.

To investigate the dependence of the RPL intensity on the radius of the disk and on the radiative lifetime of the orange and blue RPL, a line pattern (orange coloured area) was drawn on the surface of the detector as shown in Fig. 3(a). The width of the line written by X-rays was approximately 7 mm. Figure 3(b) shows the waveform displayed as a function of the radius, i.e., r=45, 30, 20, 10, 5 and 2 mm, as rotating at a speed of 2400 rpm (period=25 ms) and also shows the corresponding length per division on an oscilloscope. Figure 3(b) also shows the reconstructed dose image of the irradiated line.

As shown in Fig. 3 (b), the spatial resolution of the inner disk becomes higher than that of the outer disk. In this work, each revolution of the disk is divided into 12,500 pixels; therefore, the minimum pixel size can be estimated as 25 μm at r=50 mm and as 3.7 μm at r=7.5 mm. In addition, the transit time of the luminescence, moving across the spot size with nearly 1 μm in diameter focused by an objective lens (100×), can be estimated as 80 ns at r=50 mm and as 530 ns at r=7.5 mm. Therefore, the RPL intensity depends strongly on the radiative lifetime of the detector when the revolutions of the disk are constant (in this work, 2400 rpm).

For the lifetime of the RPL, the following experiments were performed: the decay curves of the blue and orange RPL for X-ray-irradiated glass were taken at RT. The lifetime value of the blue RPL was considerably shorter than that of the orange RPL. In addition, the lifetime values depend on the absorbed doses, when increasing the doses from 0.1 to 10 Gy, and the blue and orange RPL values decrease from 4.7 to 4.3 ns and from 2400 to 2200 ns, respectively. As a result, in the case of the blue RPL, the RPL intensity depends strongly on the lifetime (4.5 ns under a dose of 1Gy) because the intensity decreases remarkably at the inner disk.

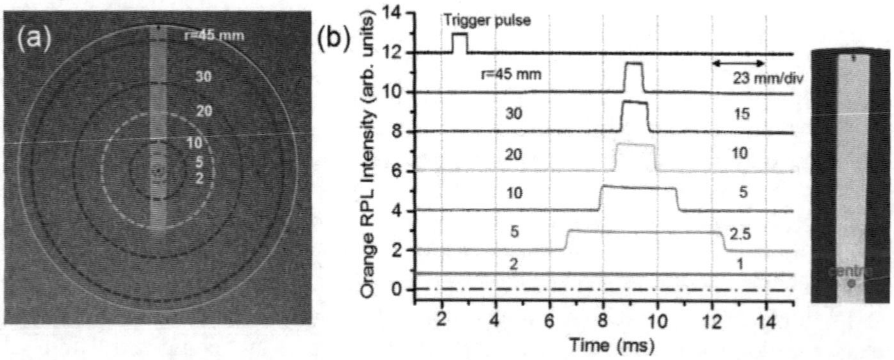

Figure 3. (a) A line pattern with a width of 7 mm drawn on the surface by X-rays. (b) Waveform of the orange RPL intensity as a function of the radius at various positions and the reconstructed 2D image (right).

Figure 4 (a) shows a photo image ("Hassam House" in Kobe) of the mask used for X-ray irradiation. Figures 4 (b) and (c) show a set of reconstructed dose distributions obtained from the disk-type Ag-activated phosphate glass based on the orange and blue RPL, respectively, with an absorbed dose of 3 Gy. The full-scale input voltage of an A/D converter was set between 0 and 2.5 V with a resolution of 16 bit. The reconstructed images indicate the red regions of the colour axes corresponding to higher concentration levels, while the blue regions indicate areas in which soft X-rays could not pass through the 250-μm-thick stainless steel attached to the detector as a mask or as non-exposure parts during X-ray irradiation. Figure 4 (d) shows an enlargement map of the parts in Fig. 4 (b). In the case of the RPL detector, steep gradients of dose distribution levels are clearly resolved with the highest image contrast. Changing the full-scale input voltage after acquiring a reconstructed 2D image, lower level distributions can be observed as low as several mGy.

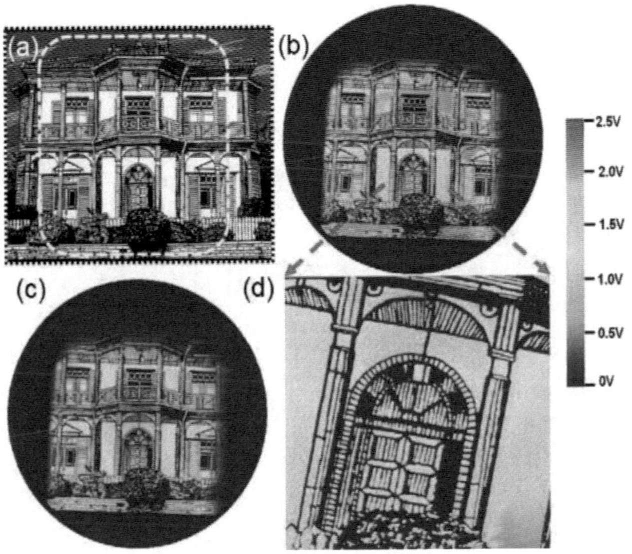

Figure 4. (a) Photo image of the mask used for X-ray irradiation. A set of reconstructed dose distributions acquired using the (b) orange and (c) blue RPL, respectively, and (d) an enlargement image of (b).

Figure 5 shows a series of the 2D reconstructed dose distributions obtained from the disk-type Ag-activated glass detector with a dose value of 3 Gy. The use of a confocal detection system and a 0.90 NA objective lens having a working distance of 1 mm enables one to reconstruct 3D dose distributions within the transparent detector by combining area images at different depths, i.e., the vicinity of the surface, 100, 200, 300 and 400 μm below the surface.

Figure 5. Reconstructed dose distributions acquired for each image at different depths of 0, 100, 200, 300 and 400 μm below the surface.

4. Conclusion

The data obtained in this study led to the following conclusions: (1) The capabilities of the reconstructed 2D dose images based on the orange and blue RPL in a disk-type Ag-activated phosphate glass with a high spatial resolution on the order of microns, a wide dynamic range covering three orders of magnitude and a non-destructive readout were successfully demonstrated. (2) The spatial resolution and RPL intensity of a disk-type detector depend strongly on the radius of the disk and the lifetime values of the material, respectively. (3) The use of a confocal detection system and the high luminescent RPL glass allow one to reconstruct a 3D image by combining each image at different depths from the surface to ≈400 μm.

Three processes, including preheating, reading and erasing, which are necessary during the use of the detector, can be performed with only a UV laser by adjusting the stepwise output levels and are currently underway. In addition, the use of the blue RPL allows measurements of the high-speed revolutions of the disk. This detector should be suitable for microradiography, X-ray microscopy, radiation diagnostics and fluorescent nuclear track applications.

Acknowledgements

T. K would like to thank Mr. Y. Koguchi and Mr. N. Takeuchi at Chiyoda Technol Corporation for their contributions to the sample preparation. This work was supported by JSPS, Grand-in-Aid for Scientific Research (B), No.26289362.

References

[1] Yamamoto T, Maki D, Sato F, Miyamoto Y, Nanto H and Iida T 2011 *Radiat. Meas.* ***46*** *1554.*
[2] Yokota R and Imagawa R 1966 *J. Phys. Soc. Jpn.* ***23*** *1038.*
[3] Perry J A 1987 *Radiophotoluminescence in Health Physics (Adam Hilger) New York.*
[4] Yukihara E G and McKeever S W S 2011 *Optically Stimulated Luminescence (John Wiley & Sons) New York.*
[5] Akselrod G M, Akselrod M S, Benton E R and Yasuda N 2006 *Nucl. Instr. and Meth. B* ***247*** *295.*
[6] Baldacchini G, Bonfigli F, Faenov A, Flora F, Montereali R M, Pace A, Pikuz T, Reale L 2003 *J. Nanosci. Nanotech.* ***3*** *483.*
[7] Montereali R M, Almaviva S, Bonfigli F, Cricenti A, Faenov A, Flora F, Caudio P, Lai A, Martellucci S, Nichelatti E, Pikuz T, Reale L, Richetta M, Vincenti M A 2010 *Nucl. Instrum. Methods Phys. Res. Sect. A* ***623*** *758.*
[8] Kurobori T, Miyamoto Y, Maruyama Y, Yamamoto T and Sasaki T 2014 *Nucl Instrum. Methods Phys. Res. Sect. B* ***326*** *76.*
[9] Olko P 2010 *Radiat. Meas.* ***45*** *506.*
[10] Garcier Y, Cordier G, Pauron C and Fazileabasse J 2007 *Radiat. Prot. Dosim.* ***124*** *107.*
[11] Knežević Ž, Stolarczyk L, Bessieres I, Bordy J M, Miljannić S and Olko P 2013 *Radiat. Meas.* ***57*** *9.*
[12] T Hsu S -M, Yeh S -H, Lin M -S and Chen W -L 2006 *Radiat. Prot. Dosim.* ***119*** *327.*
[13] Kurobori T and Nakamura S 2012 *Radiat. Meas.* ***47*** *1009.*
[14] Kurobori T and Matoba A 2014 *Jpn. J. Appl. Phys.* ***53*** *02BD14.*
[15] Kurobori T, Zheng W, Miyamoto Y, Nanto H and Yamamoto T 2010 *Opt. Mater.* ***32*** *1231.*
[16] Zheng W and Kurobori T 2011 *J. Lumin.* ***131*** *36.*

Density functional simulation of carbon at the titanium site in perovskite barium titanate

Meaad S AL-Hadidi[1,2], J P Goss[1], P R Briddon[1], Raied A Al-Hamadany[1,3], Mariam E Ahmed[4] and M J Rayson[5]

[1]Electrical and Electronic Engineering, Newcastle University, Newcastle upon Tyne, UK
[2]Renewable Energy Department, Science College, Mosul University, Mosul, Iraq
[3]Physics Department, Science College, Thi-Qar University, Nasiriyah, Iraq
[4]Physics Department, Tripoli University, Tripoli, Libya
[5]Department of Chemistry, University of Surrey, Guildford, Surrey GU2 7XH, United Kingdom

E-mail: m.al-hadidi@ncl.ac.uk

Abstract. The perovskite family includes many titanates which have been used in various applications. $BaTiO_3$ is interesting due to its room-temperature ferroelectric properties and relative low toxicity. Organic precursors present during growth make carbon a potentially key impurity, which would subsequently impact upon the $BaTiO_3$ properties. This paper presents a density function study of the structural and electronic properties of carbon substituting for Ti in rhombohedral $BaTiO_3$. The local vibrational modes of the defect centre has been calculated and suggested as a possible route to experimental identification.

1. Introduction

$BaTiO_3$ is a ferroelectric material with a high dielectric constant and low loss characteristics, widely utilised in the manufacture of electronic components such as multilayer dielectric capacitors and high refractive index thin films[1]. Its piezoelectric and pyroelectric properties lend it to passive infrared detectors[2], piezoelectric actuators and sonar devices[3], and its positive temperature coefficient of resistivity allows it to be used in sensor applications[3].

In recent years, significant amount of experimental and theoretical study has been directed at ferroelectric perovskite $BaTiO_3$[4, 5, 6, 7, 8, 9, 10, 11, 12, 13, 14]. Both theoretical and experimental studies have shown that intrinsic and extrinsic defects affect the electronic structure, photonic properties, remnant polarisation, motion of domain walls, dielectric constant and leakage current in this material[15, 16, 17, 18]. For example, pentavalent dopants (Sb, Nb and Ta) can produce semiconductor by substitution at Ti^{+4} sites[1], and oxygen vacancies also act electrically[19]. Indeed, addition of impurities to $BaTiO_3$ is needed to achieve the desired enhancement of, for example, photoconductivity, piezoelectricity and positive temperature coefficient of resistivity[20, 3, 21].

Given the dependence of the $BaTiO_3$ properties upon both stoichiometry and purity, the crystal growth techniques used are highly significant. Various different techniques have been used to synthesis thin films, including solid-state reaction[22], the sol-gel method[23, 24, 25, 16, 19], mechanochemical synthesis[26, 27], solvothermal synthesis methods[28], sputter deposition[29], pulsed-laser deposition[30], chemical vapour deposition[31] and atomic layer deposition[31, 32], in order to eliminate problems such as heterogeneity, impurities, inaccurate thickness control and the poor electrical properties of sintered ceramics. In several of these methods organic precursors are used for Ba sources, and subsequent carbon contamination has been reported in $BaTiO_3$ thin films[33, 25, 26, 32]. For example, sharp peaks at 1425, 1050, 860, and 695 cm^{-1} were observed and attributed to the carbonate ions (CO_3^{-2})[34].

In this paper, we present the results of first principles calculations on the geometry, electronic structure, vibrational modes and reorientation barrier of carbon substituting for Ti in $BaTiO_3$.

2. Computational details

The properties of carbon substituting for Ti in rhombohedral $BaTiO_3$ are calculated using first-principles density functional theory within the local density approximation[35]. The self-consistence total energy calculations were performed using the AIMPRO computer code[36, 37] based on the stable and efficient preconditioned conjugate gradients method with the optimised structures having forces of $< 10^{-3}$ atomic units and the final structural optimisation step is required to result in a reduction in the total energy of less than 10^{-5} Ha.

Atoms are modelled using norm-conserving, separable pseudo potentials[38] (valance sets of Ba, Ti, O and C are $5s^2 5p^6 6s^2$, $3s^2 3p^6 3d^2 4s^2$, $2s^2 2p^4$ and $2s^2 2p^2$, respectively), with the Kohn-Sham eigen-functions expanded using atom centred Gaussian basis sets[39], comprised from four independent sets of s, p and d type functions. Matrix elements of the Hamiltonian are determined using a plane-wave expansion of the density and Kohn-Sham potential[40] with a cut-off of 150 Ha, resulting in absolute convergence of the total energy with respect to the expansion of the charge density to better than 0.03 meV.

For bulk rhombohedral $BaTiO_3$ symmetry, our computational approach yields an equilibrium lattice parameters of 3.97 Å and 89.85°, within 1% of experiment[41, 42]. The calculated bandgap is 2.3 eV, reflecting the well documented underestimate arising from the underpinning methodology. Our estimate is consistent with previous comparable calculations[43].

To model defects, we use the same type of cell detailed previously[44], consisting of 160 host atoms. The Brillouin zone is sampled using a uniform $2 \times 2 \times 2$ Monkhorst-Pack mesh[45].

Vibrational modes have been calculated by obtaining the second derivatives of the energy with respect to atomic positions, which are then assembled into the dynamical matrix. The derivatives are determined using a finite-difference approximation based upon analytical forces obtained with atomic displacement of 0.2 atomic units.

Finally, reaction pathways and activation energies are determined using the climbing nudged-elastic-band (NEB) method[46, 47], and the convergence of the saddle point energy with respect to the number of images and the image-forces has been established to within a few meV.

3. Results and discussion

3.1. Defect geometry

We start with the geometry of carbon substitution of titanium, C_{Ti}, which may be anticipated on the grounds of both C and Ti being group-IV elements[48]. The first structure analysed was created from the octahedral symmetry, with a Ti ion simply replaced with carbon. In the absence of any symmetry constraint, the optimised structure of C_{Ti} exhibits significant structural change from the initial geometry. The carbon is displaced from the centre where it was initially co-ordinated with six oxygen ions, toward the centre of a triangular group composed of three oxygen ions, as shown in figure 1(b). The final structure has three C–O bonds 35% shorter than the host Ti–O distances. The equilibrium geometry can be clearly identified as a carbonate group ion. To further support this view it is noted that the calculated C-O bond-lengths of 1.29 Å are close to those of the carbonate group[49, 50] and those in witherite[51].

We now turn to the impact that the formation of the carbonate group has upon the electronic properties of the material.

3.2. Electronic structure

The band structure for C_{Ti} is shown in figure 1(c). The inclusion of a carbon atom at a Ti site does not have any significant effect on the band-gap, which we note is in contrast to the situation with C_{Ti} in cubic $SrTiO_3$[48]. Carbon substitution for titanium in $BaTiO_3$ is therefore expected to lead to electrically inactive defects, i.e. these centres are not expected to introduce any states in the band-gap that would lead to a change of charge state.

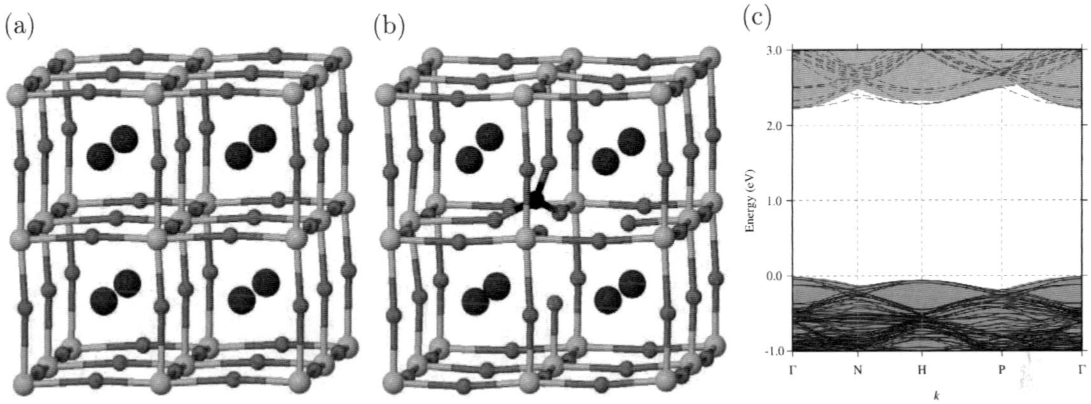

Figure 1. Schematics of (a) pure $BaTiO_3$, and (b) C_{Ti} in $BaTiO_3$. Blue, grey, red and black atoms represent Ba, Ti, O and C respectively. Vertical and horizontal axes are approximately [001] and [010], respectively, with the tilted view adopted to aid clarity. (c) shows the band structure of C_{Ti} in $BaTiO_3$ in the first Brillouin zone of the supercell, plotted along high-symmetry branches. Labelling on the x-axis follows the conventions for special points at the zone-boundary for the bcc Brillouin zone. Occupied and empty bands are shown in solid blue and dashed red lines, respectively. The underlying shaded areas show the regions of the bands for the corresponding C-free supercell. The zero on the energy scale is defined to be the highest occupied state in each case.

3.3. Vibrational modes

Turning to vibrational characteristics, we find that the calculated frequencies are consistent with C_{Ti} resulting in the formation of carbonate ions. A set of characteristic local vibrational modes lies above the one-phonon maximum, which could prove highly effective in the experimental identification of this centre: with a nearly degenerate C–O stretch mode at 1383 cm^{-1} and 1379 cm^{-1}, and a breathing stretch-mode is at 1061 cm^{-1}. These are close to the experimental values of CO_3^{-2} gas phase[52] which have E' and A_1' stretch modes lying around 1415 cm^{-1} and 1064cm, respectively, and also agree both with the modes detected in $BaTiO_3$[34], and with those modes assigned to the carbonate groups in $BaCO_3$[53]. Lower frequency modes of the carbonate group are also present for C_{Ti}, but are resonant with the one-phonon density of states. All three local modes listed here are, by symmetry, infra-red and Raman active, and given sufficient concentrations one might seek to confirm the presence of this form of carbon centre directly via either IR or Raman spectroscopy.

3.4. Carbonate group reorientation

Finally, the energetics for carbonate group reorientation have been predicted. The initial and final structures are shown in figure 2(a) and (b), and the energy profile along the minimum energy path between them in figure 2(c). The energy difference between initial and final structure has been calculated to be about 50 meV, which is related to non-equivalent in the defect with respect to the rhombohedral distortion. The activation energy for reorientation process has been calculated to be around 2.5 eV. The barrier is found to be high in energy compared to, for example, diffusion of the oxygen vacancy, since the reorientation requires the breaking and formation of strong covalent C–O bonds. It is noted, however, that this activation energy is similar to the equivalent path in cubic $SrTiO_3$[48]. The high barrier is expected to lead to slow reorientation at room temperature, which will impact upon the hysteresis seen in the material under an applied electric field.

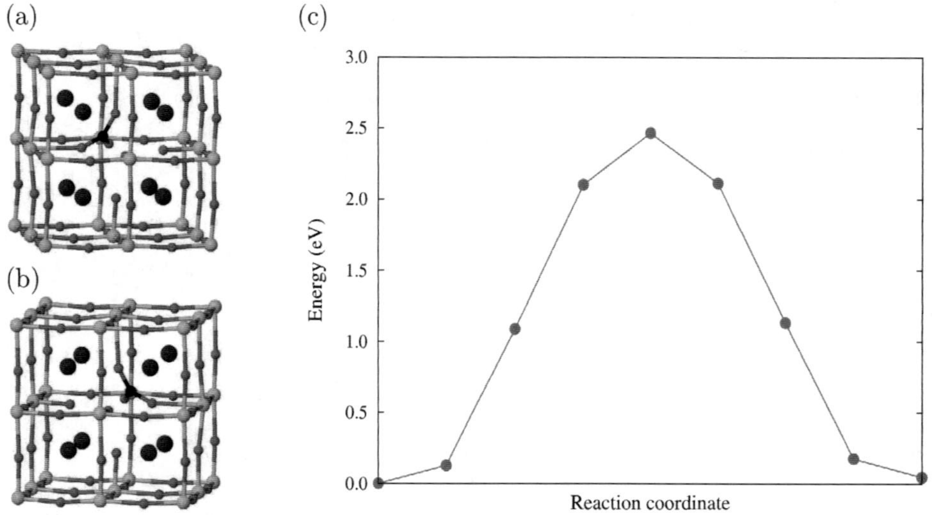

Figure 2. Schematic of (a) initial, and (b) final structure of C_{Ti} in $BaTiO_3$ in a simple reorientation process. Colours and orientation are as indicated in figure 1. (c) Calculated reorientation barrier of C_{Ti} in $BaTiO_3$ between initial and final structure, with circles showing the energies of the 9 images in the NEB algorithm.

4. Summary and conclusions

In summary, based on first-principles density-functional calculations, the characteristics of carbon impurities substituting for titanium in rhombohedral $BaTiO_3$ have been examined, including the structural configuration, electronic properties, vibrational modes and reorientation barrier. The calculations reveal that carbon energetically favours the formation of carbonate groups and suggest that the carbon substituting of titanium is electrically passive, so that C_{Ti} is predicted to have a minimal impact upon the electrical conductivity of $BaTiO_3$. The vibrational frequencies associated with defect centre have been determined and compared with available theoretical and experiment values. These spectroscopic properties are a clear route to experimental identification of this defect. Finally, the reorientation barrier for the carbonate group for C_{Ti} has been estimated, and because the process involves the breaking and forming of C–O covalent bonds, the barrier has a relatively high value which will in general impact upon ferroelectric hysteresis.

References

[1] Chatterjee S, Stojanovic BD and Maiti H, 2003 *Mater. Chem. Phys.* **78** 702
[2] Akcay G, Zhong S, Alpay S and Mantese J, 2006 *Solid State Commun.* **137** 589
[3] Ertug B, 2013 *American Journal of Engineering Research* **2** 1
[4] Ishidate T, Abe S, Takahashi H and Môri N, 1997 *Phys. Rev. Lett.* **78** 2397

[5] Piskunov S, Heifets E, Eglitis R and Borstel G, 2004 *Comp. Mater. Sci.* **29** 165–178
[6] Stuhrmann HB, 1970 *Acta Crystallogr. A* **26** 297
[7] Kinoshita K and Yamaji A, 1976 *J. Appl. Phys.* **47** 371
[8] Frey MH and Payne DA, 1996 *Phys. Rev. B* **54** 3158
[9] King-Smith RD and Vanderbilt D, 1994 *Phys. Rev. B* **49** 5828
[10] Saha S, Sinha TP and Mookerjee A, 2000 *Phys. Rev. B* **62** 8828
[11] Roussev R and Millis AJ, 2003 *Phys. Rev. B* **67** 014105
[12] Zhong W, Vanderbilt D and Rabe KM, 1994 *Phys. Rev. Lett.* **73** 1861
[13] Tanaka T, Matsunaga K, Ikuhara Y and Yamamoto T, 2003 *Phys. Rev. B* **68** 205213
[14] Miura K, Azuma M and Funakubo H, 2011 *Materials* **4** 260
[15] Pöykkö S and Chadi DJ, 1999 *Phys. Rev. Lett.* **83** 1231
[16] Ghosh VJ, Nielsen B and Friessnegg T, 2000 *Phys. Rev. B* **61** 207
[17] Schwartz RN and Wechsler BA, 2000 *Phys. Rev. B* **61** 8141
[18] Stashans A and Chimborazo J, 2002 *Phil. Mag. B* **82** 1145
[19] Yoo HI, Chang MW, Oh TS, Lee CE and Becker KD, 2007 *J. Appl. Phys.* **102** 093701
[20] Chen X, Lu W and Shen SC, 2004 *Solid State Commun.* **130** 641
[21] Rao M, Ramesh KaRM and Rao B, 2013 *Adv. in Mater. Phys. and Chem.* **87** 77
[22] Potdar H, Deshpande S and Date S, 1999 *Mater. Chem. Phys.* **58** 121
[23] Cho W, 1998 *J. Phys. Chem. Solids* **59** 659
[24] Cho W and Hamada E, 1998 *J. Alloys and Compounds* **266** 118
[25] Li B, Wang X and Li L, 2003 *Mater. Chem. Phys.* **78** 292
[26] Stojanovic B, Simoes A, Paiva-Santos C, Jovalekic C, Mitic V and Varela J, 2005 *J. Euro. Ceramic Soc.* **25** 1985
[27] Stojanovic BD, Jovalekic C, Vukotic V, Simoes AZ and Varela JA, 2005 *Ferroelectrics* **319** 65
[28] Lee H, Moon S, Choi C and Kim DK, 2012 *J. Am. Ceramic Soc.* **95** 2429
[29] Reynolds GJ, Kratzer M, Dubs M, Felzer H and Mamazza R, 2012 *Materials* **5** 575
[30] Norton MG, Scarfone C, Li J, Carter CB and Mayer JW, 1991 *J. Mater. Res.* **6** 2022
[31] Holme TP and Prinz FB, 2007 *J. Phys. Chem. A* **111** 8147
[32] Vehkamäki M, Hatanpää T, Ritala M, Leskelä M, Väyrynen S and Rauhala E, 2007 *Chem. Vapor Dep.* **13** 239
[33] Vinothini V, Singh P and Balasubramanian M, 2006 *Ceramics International* **32** 99
[34] Tsay J, Fang T, Gubiotti TA and Ying JY, 1998 *J. Mater. Sci.* **33** 3721–3727
[35] Perdew JP and Wang Y, 1992 *Phys. Rev. B* **45** 13244–13249
[36] Briddon PR and Jones R, 2000 *Phys. Status Solidi B* **217** 131–171
[37] Rayson MJ and Briddon PR, 2008 *Computer Phys. Comm.* **178** 128–134
[38] Hartwigsen C, Goedecker S and Hutter J, 1998 *Phys. Rev. B* **58** 3641–3662
[39] Goss JP, Shaw MJ and Briddon PR, 2007 In *Theory of Defects in Semiconductors*, edited by DA Drabold and SK Estreicher, volume 104 of *Topics in Applied Physics*. Springer, Berlin/Heidelberg, pages 69–94
[40] Rayson MJ and Briddon PR, 2009 *Phys. Rev. B* **80** 205104
[41] Berlincourt D and Jaffe H, 1958 *Phys. Rev.* **111** 143
[42] Hewat AW, 1973 *Ferroelectrics* **6** 215
[43] Ghosez P, Gonze X and Michenaud JP, 1999 *Ferroelectrics* **220** 1
[44] Al-Hamadany R, Goss JP, Briddon PR, Mojarad SA, Al-Hadidi M, ONeill AG and Rayson MJ, 2013 *J. Appl. Phys.* **113** 024108
[45] Monkhorst HJ and Pack JD, 1976 *Phys. Rev. B* **13** 5188–5192
[46] Henkelman G, Uberuaga BP and Jónsson H, 2000 *J. Chem. Phys.* **113** 9901–9904
[47] Henkelman G and Jónsson H, 2000 *J. Chem. Phys.* **113** 9978–9985
[48] Al-Hadidi MS, Goss JP, Briddon PR, Al-Hamadany RA and Ahmed ME, 2013 *J. Phys. Conf. Ser.* **472** 012006
[49] Adler HH and Kerr PF, 1963 *Am. Miner.* **48** 839–853
[50] Gagliardi L and Roos BO, 2002 *Inorg. Chem.* **41** 1315
[51] Antao SM and Hassan I, 2007 *Phys. and Chem. Minerals* **34** 573–580
[52] Tavender SM, Johnson SA, Balsom D, Parker wA and Bisby RH, 1999 *Laser Chem.* **19** 311–316
[53] Krishnamurti D, 1960 *Proceedings of the Indian Academy of Sciences - Section A* **51** 285–295

Multiple thermoluminescence glow peaks and afterglow suppression in CsI:Tl co-doped with Eu^{2+} or Yb^{2+}

R H Bartram[1], L A Kappers[1], D S Hamilton[1], C Brecher[2], E E Ovechkina[3], S R Miller[3] and V V Nagarkar[3]

[1]Department of Physics, University of Connecticut, Storrs, CT 06269, USA
[2]Alem Associates, 44 Hunt St., Watertown, MA 02472, USA
[3]Radiation Monitoring Devices, 44 Hunt St., Watertown, MA 02472, USA

E-mail: RHBartram2@aol.com

Abstract. CsI:Tl is a widely utilized scintillator material with many desirable properties but its applicability is limited by persistent afterglow. However, effective afterglow suppression has been achieved by co-doping with divalent lanthanides. The present report is concerned with observation of multiple thermoluminescence glow peaks in CsI:Tl,Eu and CsI:Tl,Yb, attributed to varying distributions of charge-compensating cation vacancies relative to divalent lanthanide co-dopants, and the subsequent modification of these distributions by repeated observations. It is observed that Yb^{2+} provides a slightly shallower electron trap than Eu^{2+}, and that it can occupy a face-centered position by virtue of its relatively small ionic radius; the latter observation is confirmed by electrostatic calculations. It is also found that repeated observation of thermoluminescence in these materials has a modest adverse effect on afterglow suppression.

1. Introduction

Afterglow suppression in scintillator material CsI:Tl, co-doped with various divalent rare-earth ions, has been investigated extensively, as described in several previous publications.[4,5,6,7,8,9,10,11,12] Thermoluminescence measurements[5,6] of CsI:Tl revealed glow peaks at 60K, 90K and 120K, attributable[13,14] respectively to thermally-activated tunneling recombination, the onset of V_K-center mobility, and thermal ionization of electrons from Tl^0. It has also been established that radiative recombination occurs at Tl sites, where the emitting state is (Tl^+ + STE),[6] and that Tl^{2+} is unstable in CsI.[15] In the present work, repeated measurements are found to modify the distribution of glow peaks while preserving their overall intensity.

2. Sample preparation

Single crystals 10 mm in diameter and 35 mm long, subsequently sliced into samples 2.5 mm thick, were grown by the vertical Bridgeman method at RMD. Two samples were employed in the present investigation: sample SC8-1#3 with 0.2% Tl and 0.2% Eu in the starting material, and sample SC66A-1#2 with 0.2% Tl and 0.1% Yb in the starting material. However, the proportions of dopants incorporated in the finished crystal samples varied substantially from one sample to another and within each sample.

3. Experiment.

Radioluminescence and thermoluminescence measurements were performed at the University of Connecticut on the single-crystal samples of co-doped CsI:Tl,Eu and CsI:Tl,Yb prepared at RMD. The branching ratio of electron hole pairs that contribute either to scintillation or thermoluminescence was determined by employing a common apparatus for both measurements. An electron Van de Graaff accelerator operated at a beam voltage of 1.0 MeV and a beam current of 1.0 μA was employed as a primary radiation source, with a thin copper target that served as a point source of ~0.5 MeV gamma rays. Each sample was mounted in turn on a heated pedestal monitored by a thermocouple and cooled by flowing nitrogen gas through a heat exchanger. Luminescence was conducted to a photomultiplier by a shielded optical fiber. Samples were irradiated for four minutes at reduced temperatures to record radioluminescence and then subjected to a temperature ramp of +25°C per minute at an enhanced gain setting to record thermoluminescence. These two steps were repeated alternately several times for each sample, with varying results.

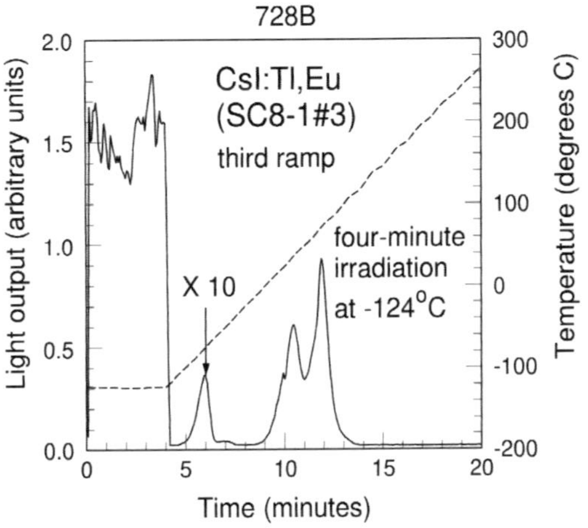

Figure 1. Light output of an as-received sample of CsI:Tl, irradiated at -63°C and subsequently heated to 280°C with a temperature ramp of +25°C/min.

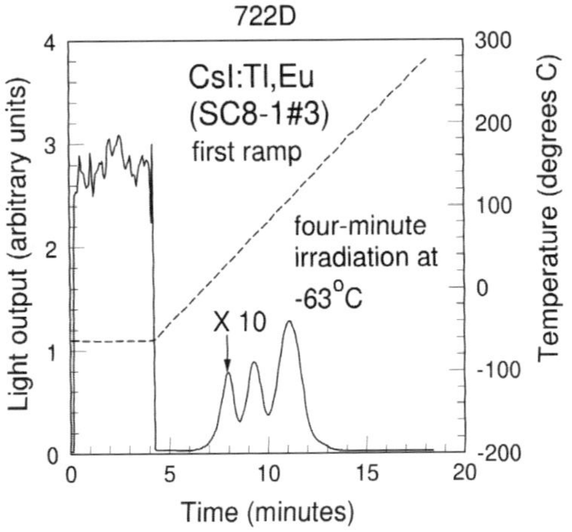

Figure 2. Light output of the same sample of CsI:Tl,Eu as in Figure 1, during the third irradiation at -124°C and the third temperature ramp of +25°C/min to 280°C.

IOP Conf. Series: Materials Science and Engineering **80** (2015) 012003 doi:10.1088/1757-899X/80/1/012003

Three distinct glow peaks are observed after the first irradiation of CsI:Tl,Eu, as shown in Figure 1. The highest-temperature peak near 100°C, which is broader than the others, probably contains multiple components. After the third irradiation, the highest temperature peak has disappeared and a new peak has appeared at lower temperature, as shown in Figure 2. A similar transformation occurs in CsI:Tl,Yb. Multiple glow peaks are attributed to the varying proximity of charge-compensating cation vacancies to the co-dopant. Presumably, cation vacancies can migrate at the elevated temperatures (≤ 280°C) achieved during thermoluminescence by changing places with neighboring cations, and the total energy of the sample is reduced by their increased proximity to divalent co-dopant cations. The effect of a nearby charge-compensating cation vacancy on the co-dopant trap depth can be approximated by Coulomb interaction of the excess charges involved in their unperturbed lattice positions, diminished by the dielectric constant of the medium. Accordingly, it can be calculated from the formula

$$\Delta E_n = -\frac{e^2 / a_0}{\kappa(a / a_0)\sqrt{(\alpha_n^2 + \beta_n^2 + \gamma_n^2)}}, \tag{1}$$

where e^2/a_0=27.2 eV, κ=6.31, a=4.56 Å, a_0=0.529 Å, and α_n, β_n and γ_n are integers denoting the position of the charge-compensating cation vacancy relative to the europium or yterbium impurity in units of the lattice parameter. The parameters κ and a are, respectively, the dielectric constant and lattice parameter of CsI. Values of ΔE_n calculated from Eq. (1) are listed in Table 1. The rate of thermal ionization in a glow peak corresponding to a trap depth E is given by the formula

$$p = s\exp(-E / k_B T), \tag{2}$$

where T is in degrees Kelvin. The value of the pre-exponential factor s determines the temperature width of the thermoluminescence glow peak. If one assumes that s is approximately the same for all traps, based on the observation that the temperature widths of the various glow peaks are comparable, it follows that the trap depths and temperatures for successive glow peaks are related by

$$\frac{\Delta E_n}{E_\infty} = \frac{\Delta T_n}{T_\infty}. \tag{3}$$

Table 1. Values of ΔE_n for CsI:Tl,Eu and CsI:Tl,Yb, calculated from Eq. (1)

n	$(\alpha_n,\beta_n,\gamma_n)$	$\frac{1}{\sqrt{(\alpha_n^2+\beta_n^2+\gamma_n^2)}}$	ΔE_n (eV)	$T_\infty+\Delta T_n$ (K) (Eu^+/Eu^{2+})	$T_\infty+\Delta T_n$ (K) (Yb^+/Yb^{2+})
1	(±1,0,0),(0,±1,0),(0,0,±1)	1.0	-0.500	233	163
2	(±1,±1,0),(±1,0,±1),(0,±1,±1)	0.707	-0.354	303	233
3	(±1,±1,±1)	0.577	-0.289	334	---
4	(±2,0,0),(0,±2,0),(0,0,±2)	0.5	-0.250	---	---
5	(±2,±1,0),(±2,0,±1),(±1,±2,0), (0,±2,±1),(±1,0,±2),(0,±1,±2)	0.447	-0.224	---	---
6	(±2,±1,±1),(±1,±2,±1),(±1,±1,±2)	0.408	-0.204	375	---

IOP Conf. Series: Materials Science and Engineering **80** (2015) 012003 doi:10.1088/1757-899X/80/1/012003

One can then infer the values of E_∞ and T_∞ in the absence of a charge-compensating cation vacancy from Eq. (3) and Table 1, which also lists the approximate temperatures of observed glow peaks. The trap depths for isolated co-dopants are inferred to be $E_\infty \approx 1.00$ eV for Eu^+/Eu^{2+} and $E_\infty \approx 0.84$ eV for Yb^+/Yb^{2+}.

4. Anomalous Thermoluminescence Glow Peak in CsI:Tl,Yb

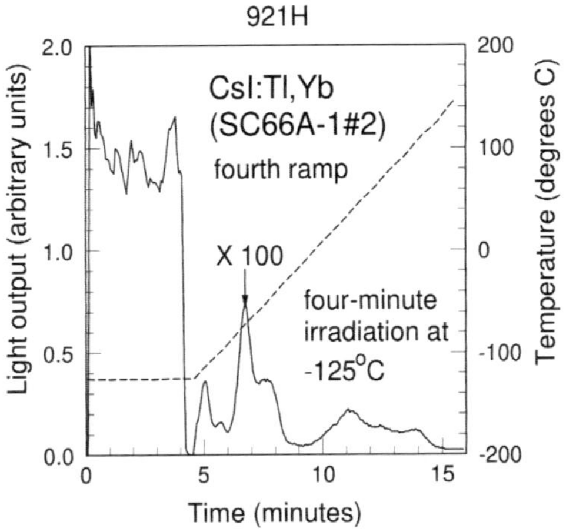

Figure 3. Light output of a sample of CsI:Tl,Yb during the fourth irradiation at -125°C and the fourth temperature ramp of +25°C/min to 150°C. The glow peak at -70°C is anomalous.

The scintillation and thermoluminescence of CsI:Tl,Yb during the fourth irradiation and temperature ramp of the present investigation are shown in Figure 3. The most prominent thermoluminescence glow peak at -70°C (203K) is anomalous in that it doesn't fit Equation (1) and Table (1). This anomalous peak is attributed to Yb^{2+} in the face-center position with a cation vacancy in the adjacent cell. That situation can occur in CsI because the ionic radius of Yb^{2+} is smaller than that of Eu^{2+} and substantially smaller than that of Cs^+. The radius of the face-center position in undistorted CsI is given by the formula

$$r_{FC} = a/\sqrt{2} - r_{I-} , \tag{4}$$

where a is the lattice parameter of CsI, $a = 4.56$ Å, and r_{I-} is the ionic radius of iodine minus with six-fold coordination, $r_{I-} = 2.20$ Å. It follows that $r_{FC} = 1.02$ Å. Note that the undistorted face-centered position in CsI can accommodate Yb^{2+} but not Eu^{2+}, and was established to be more stable than the body-centered position.

5. Hole Traps

In the present experiment, light emission is observed at temperatures as high as ~100 °C. However, it has been shown that V_k-centers trapped at Tl^+ ions or cation vacancies become unstable at much lower temperatures.[16] Accordingly, it is inferred that two or more cation vacancies must stabilize a trapped V_k-center pending its combination with an electron ionized from a divalent co-dopant and the subsequent radiative recombination of the resulting exciton at a thallium site.

References

[4] Nagarkar V V, Ovechkina E E, Miller S R, Gaysinskiy V, Brecher C, Lempicki A and Bartram R H 2005 "Luminescence properties of CsI:Tl crystals codoped with Eu", *Functional Materials* **12** 645-651

[5] Brecher C, Lempicki A, Miller S R, Glodo J, Ovechkina E E, Gaysinskiy V, Nagarkar V V and Bartram R H 2006 "Suppression of afterglow in CsI:Tl by codoping with Eu^{2+}. I. Experimental", *Nuclear Instruments and Methods in Physics Research* A **558** 450-457

[6] Bartram R H, Kappers L A, Hamilton D S, Lempicki A, Brecher C, Glodo J, Gaysinskiy V and Ovechkina E E 2006 "Suppression of afterglow in CsI:Tl by codoping with Eu^{2+}. II. Theoretical model", *Nuclear Instruments and Methods in Physics Research* A **558**, 458-467

[7] Kappers L A, Bartram R H, Hamilton D S, Brecher C, Lempicki A, Gaysinskiy V, Ovechkina E E and Nagarkar V V 2007 "Effect of Eu^{2+} concentration on afterglow suppression in CaI:Tl,Eu", *Radiation Measurements* **42**, 537-540

[8] Bartram R H, Kappers L A, Hamilton D S, Lempicki A, Brecher C, Gaysinskiy V, Ovechkina E E and Nagarkar V V 2008 "Afterglow suppression and non-radiative charge-transfer in CsI:Tl,Sm", *IEEE Transactions on Nuclear Science* **55**, 1232-1236

[9] Nagarkar V V, Ovechkina E E, Gaysinskiy V, Miller S R, Brecher C, Lempicki A and Bartram R H 2008 "Scintillation properties of scintillation crystals codoped with Sm^{2+}", *IEEE Transactions on Nuclear Science* **55**, 1270-1274

[10] Kappers L A, Bartram R H, Hamilton D S, Lempicki A, Brecher C, Gaysinskiy V, Ovechkina E E and Nagarkar V V 2010 "Concentration dependence of afterglow suppression in CsI:Tl,Sm", *Journal of Physics: Conference Series* **249**, 012014

[11] Kappers L A, Bartram R H, Hamilton D S, Lempicki A, Brecher C, Gaysinskiy V, Ovechkina, E E, Thacker S and Nagarkar V V 2010 "A tunneling model for afterglow suppression in CsI:Tl,Sm scintillation materials", *Radiation Measurements* **45**, 426-428

[12] Kappers L A, Bartram R H, Hamilton D S, Lempicki A, Brecher C, Gaysinskiy V, Ovechkina E E and Nagarkar V V 2010 "Scintillation, afterglow and thermoluminescence of CsI:Tl,Sm", *IOP Conference Series: Mat. Sci. Eng.* **15**, 012091

[13] Sidler T, Pellaux J P, Nouailhat A and Aegerter M A, 1973 *Solid State Communications* **13**, 479

[14] Barland M, Duval E and Nouailhat A 1981 *Journal of Physics C* **14**, 4237

[15] Rodnyi P A 1997 "Physical Processes in Inorganic Scintillators", *CRC Press*, Boca Raton, New York

[16] Yakovlev V, Trefilova L and Meleshko A 2009 "Time-resolved optical spectroscopy of CsI(Tl) crystals by pulsed electron-beam irradiation", *Journal of Luminescence* **129**, 790-796

Anisotropic distortions in Sb doped $CePt_4Ge_{12}$

F Bridges[1], P Nast[1], J Wilde[1], T Keiber[1], M B Maple[2], K Huang[3], and B D White[2]

[1] Physics Department, UCSC, Santa Cruz, CA 95064 USA
[2] Physics Department, UCSD, La Jolla, CA 92093 USA
[3] Materials Science and Engineering Program, UCSD, La Jolla, CA 92093 USA

E-mail: bridges@ucsc.edu

Abstract.
We investigated the local structure of the Sb doped skutterudite $CePt_4Ge_{12-x}Sb_x$ using the Extended X-ray Absorption Fine Structure Technique (EXAFS). As the concentration of Sb is increased the disorder around Ce increases rapidly, and for $x = 3$, the peak for the nearest neighbor (Ce-Ge) is no longer observed. In contrast, for the Pt site, the disorder of the nearest neighbors is small even for $x = 3$. Thus the distortions are anisotropic and appear to be mainly in the plane of the Ce-Ge bonds and Ge_4 rings. The increased disorder about Ce will decrease the lattice thermal conductivity at low temperatures, and likely is part of the reason for improved thermoelectric properties for the $x = 1$ sample.

1. Introduction
The filled skutterudites LnX_4T_{12} (Ln = rare earth; X = Fe, Ru, Os, Pt; T = P, As, Sb, Ge) are a broad class of materials with several unusual properties[1], including a low thermal conductivity, κ, and a high Seebeck coefficient, both of which are important for thermoelectric applications. The compound $CePt_4Ge_{12}$, has only recently been synthezised[2, 3, 4, 5] and no local structure studies have been reported. For this material (space group $Im\bar{3}$), twenty-seven Pt atoms form a cubic unit cell composed of eight smaller cubic cages. Six of these cages contain square rings of four Ge atoms, while the other two cages enclose Ce atoms. The Ln atoms in other skutterudites exhibit "rattling" behavior, with large amplitude vibrations and the ability to scatter phonons effectively, which contribute to the low value of thermal conductivity. Interestingly, substituting Sb for Ge greatly enhances the thermoelectric properties of the material[3]. Here we use the Extended X-ray Absorption Fine Structure (EXAFS) technique to probe the local structure in $CePt_4Ge_{12-x}Sb_x$; in particular, we investigate the rattling behavior of Ce and observe how the local structure varies with Sb doping[4] .

2. Experimental Details
EXAFS samples were prepared by grinding polycrystalline samples, prepared as in Ref. [5], into a powder, which was then filtered through a mesh sieve to produce a sufficiently fine grain. The fine powder was brushed onto scotch tape in a single layer; two pieces were pressed together to fully enclose the powder sample. These double layers were cut into strips and stacked together to achieve an absorption step height in the range 0.3 to 0.7 for the edge being studied.

EXAFS data were collected on beamlines 4-1 and 7-3 at the Stanford Synchrotron Radiation Lightsource (SSRL) using Si 220 monochromator crystals. The monochromator was detuned 50% to minimize harmonics. EXAFS transmission data at the Pt L_{III}, Ge K, and Sb K edges and fluorescence data for the Ce L_{III} edge were collected as a function of temperature T (5-300 K). EXAFS data reduction and analysis were carried out using the RSXAP package[6].

3. EXAFS Data

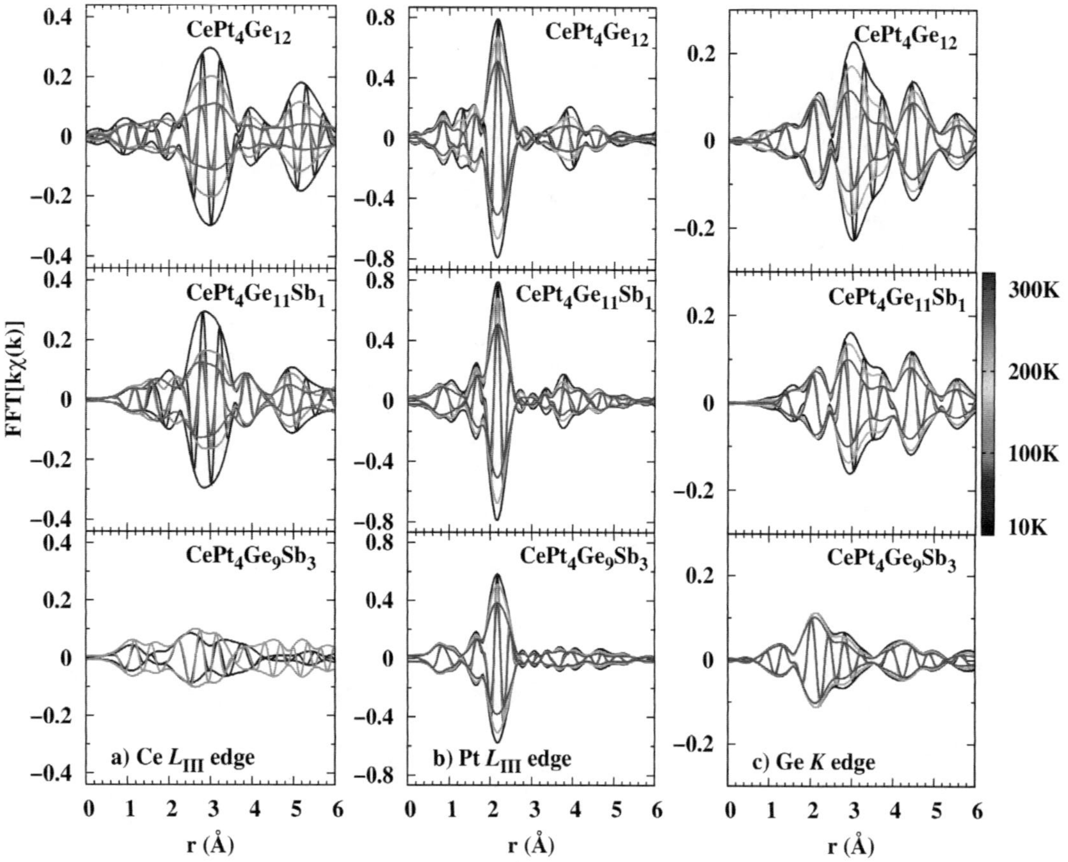

Figure 1. r-space data for a) the Ce L_{III} , b) the Pt L_{III}, and c) Ge K edge, at 8 (blue), 150 (green), and 300 K (red), for x = 0 (top), 1 (middle), and 3 (bottom). FFT ranges are 2.8-9.5, 3-14.7, and 2.9-9.7 Å for Ce, Pt, and Ge respectively.

In Figure 1 we plot the r-space data as a function of T for the Ce L_{III}, Pt L_{III}, and Ge K edges respectively for x= 0, 1, and 3. Peaks in EXAFS data are always shifted to lower r compared to actual distances from crystallographic data. For the Ce data, the first peak near 2.9 Å is a composite of the Ce-Ge peak (12 Ge neighbors at 3.32 Å) and a weaker Ce-Pt peak near 3.4 Å (8 Pt neighbors at 3.7 Å) that overlaps the Ce-Ge pair. The data for $x = 0$ or 1 show a large amplitude at 8 K and a very strong T-dependence, which is characteristic of a rattler atom. However, as more Sb is substituted, the amplitude is drastically reduced, with no clear peak or T-dependence observed for $x = 3$; the k-space data above 150 K have a low amplitude and are too noisy to reduce. Thus the environment about Ce is highly disordered even at 5 K.

For the Pt data, the first peak (2.2 Å) is a Pt-Ge pair at 2.48 Å. The group of peaks from about 3.2 to 4.5 Å is a composite of a second Pt-Ge pair at 4.14 Å, a Pt-Pt pair at 4.31 Å, as well as the weak Pt-Ce pair at 3.73 Å. The Pt data show a weaker temperature dependence, as the amplitude at 300 K is only diminished by about 35%. The data are also less affected by the doping level; even for $x = 3$, the amplitude of the first peak (Pt-Ge) at low-T is only reduced by 25%. However, the further neighbor peaks are no longer visible indicating significant distortion within a unit cell.

For the Ge data, the first (double) peak (1.8-3.5 Å) is the sum of two interfering pairs, Ge-Pt ($r = 2.48$ Å) and Ge-Ge ($r = 2.56$ Å). The second main peak from about 3.5 to 5 Å is composed of a Ge-Ge neighbor at 3.26 Å, a small Ge-Ce at 3.32 Å, and two more unresolved Ge-Ge pairs (3.62 and 3.75 Å respectively). The amplitude drops by about 50% for $x = 3$, making it more disordered than the Pt data, but less than the Ce data. An interesting phenomena is that at high doping levels, the amplitude at 2.1 Å changes little while that near 3 Å is sharply reduced, suggesting that the Ge-Ge peak disorders faster with doping than the Ge-Pt peak. Also, the interference dip at 2.4 Å disappears - from the changing interference between the first two pairs.

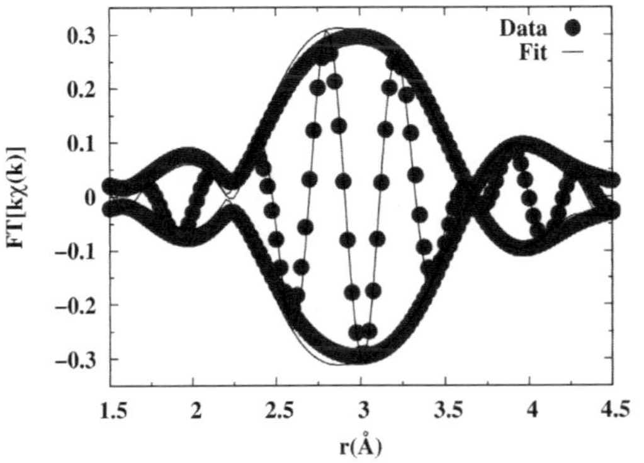

Figure 2. Example of the fit (solid black line) for the first two pairs in the Ce L_{III} edge r-space data (blue circles), collected at 8 K.

4. Analysis

The r-space data were fit to a sum of theoretical EXAFS functions calculated using the program FEFF8[7] and the known structure, taken from diffraction data[2]. For Ce, two peaks were fit (Ce-Ge and Ce-Pt) and the ratio of the number of neighbors was constrained to the known coordinations from diffraction data. From such fits, the pair distance and broadening, σ, were extracted for each pair distribution function. An example of a fit for the 8 K Ce data is shown in Figure 2 (Fit ranges: 2.0-4.4 Å for Ce data and 1.9-4.2 Å for Pt data).

From fits at multiple temperatures, σ^2(T) was extracted for two neighbors in the Ce data and three neighbors in the Pt data at $x = 0$ and 1. These are plotted in Figure 3a,b respectively. For the nearest neighbor pair, Ce-Ge, at $x = 0$, σ^2(T) follows an Einstein model well (rigid cage approximation), with a low static off-set. The $x = 1$ sample shows a higher degree of disorder, and a much larger value of σ^2_{static}. The Ce-Pt pair has a significant static contribution even at $x = 0$, and at $x = 1$ the data are too disordered to fit using an Einstein model.

The Pt σ^2(T) data are better described using a correlated-Debye model. Good fits are obtained for the Pt-Ge and Pt-Pt pairs, for $x = 0$ and 1. However, the $x = 1$ sample has a higher static offset than the undoped material.

5. Conclusions

The Einstein temperature for the nearest neighbor Ce-Ge pairs is low, consistent with rattler behavior for rare earth atoms in other skutterudites. In addition, this pair is quite well ordered at low T for $x = 0$; i.e., σ^2_{static} is small. With increasing Sb doping levels, the Ce L_{III} edge data rapidly becomes disordered (σ^2_{static} increases dramatically) and at $x = 3$, the Ce-Ge peak

IOP Conf. Series: Materials Science and Engineering **80** (2015) 012004 doi:10.1088/1757-899X/80/1/012004

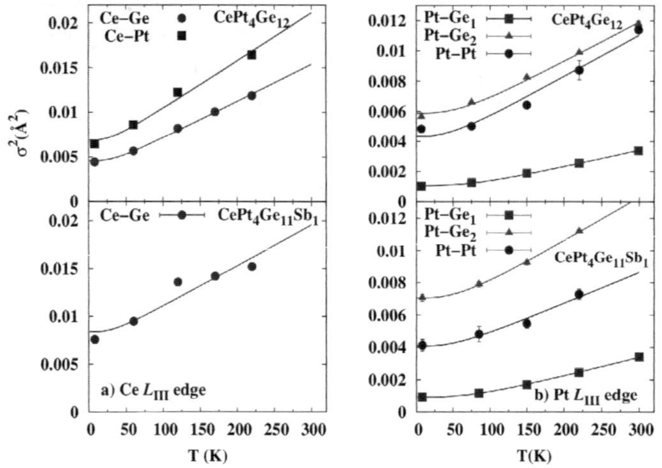

Figure 3. a) Plots of σ^2(T) for the Ce-Ge (red circles) and Ce-Pt (blue squares) for $x = 0$ (top panel) and $x = 1$ (bottom panel) samples. b) Plots of σ^2(T) for the first two Pt-Ge pairs (red squares and blue circles) and Pt-Pt (violet triangles) for $x = 0$ (top panel) and $x = 1$ (bottom panel).

Pair (Sb_0)	θ	σ^2_{static}	Pair(Sb_1)	θ	σ^2_{static}
Ce-Ge	90.7	0.00266	Ce-Ge	89.2	0.00647
Ce-Pt	79.8	0.00477			
Pt-Ge_1	357.9	-0.00037	Pt-Ge_1	350.7	-0.00054
Pt-Ge_2	282.3	0.00345	Pt-Ge_2	277.9	0.004
Pt-Pt	204.3	0.00251	Pt-Pt	241.1	0.00252

Table 1. Einstein (Ce) or correlated-Debye (Pt) temperatures (θ) and static offsets (σ^2_{static}) for several pairs at x = 0 and 1. Relative errors, ~ 2%, are small compared to systematic errors, ~ 10%.

is no longer well defined. Even at $x = 1$, the second neighbor Ce-Pt pair is too disordered to be fit to an Einstein model. In contrast, the Pt-Ge peak is only slightly distorted even at x = 3. The environment about Ge is in between; more disordered than around the Pt site but less than about the Ce site. This suggests that the distortions are primarily in the plane of the Ge_4 rings, mainly towards the rattler atom, Ce. One possibility is that the presence of the dopant pushes some Ce atoms off-center but that requires further investigation. The increased disorder will decrease the lattice thermal conductivity and likely plays a role in the improvement of thermoelectric properties for the Sb doped materials.

Acknowledgments
This work was supported under NSF grant DMR1005568. The experiments were performed at SSRL, operated by the DOE, Division of Chemical Sciences. Sample synthesis and characterization at UCSD were supported by the U. S. Department of Energy, Office of Basic Energy Sciences, Division of Materials Sciences and Engineering under Award Grant No. DE-FG02-04-ER46105.

References
[1] Sales B C 2003 in *Handbook on the Physics and Chemistry of the Rare Earths* vol 33 Chap 211 'ed L. Eyring *et al*' (Elsevier Science Publishing Co., New York) pp 1–34
[2] Toda M, Sugawara H, Magishi K-i, Saito T, Koyama K, Aoki Y, Sato H 2008 *J. Phys. Soc. Jap.* **77** 124702
[3] Nicklas M, *et al.* 2012 *Phys. Rev. Lett.* **109** 236405
[4] White B D, Huang K and Maple M B 2014 (submitted to *Phys. Rev. Lett.*)
[5] Huang K, *et al.* 2014 *Phys. Rev.* B **89** 035145
[6] See http://lise.lbl.gov/RSPAK.
[7] Ankudinov A L, Ravel B, Rehr J J and Conradson S D 1998 *Phys. Rev.* B **58** 7565

X-Ray Absorption Spectroscopy and Computer Modelling Study of Nanocrystalline Binary Alkaline Earth Fluorides

A.V. Chadwick[1], A. Düvel[2], P. Heitjans[2], D.M. Pickup[1], S. Ramos[1], D.C. Sayle[1] and T.X.T. Sayle[1]

1 School of Physical Sciences, University of Kent, Canterbury, Kent CT2 7NH
2 Institute of Physical Chemistry and Electrochemistry, Gottfried Wilhelm Leibniz University Hannover, Callinstrasse 3a, D-30167 Hannover, Germany

E-mail: a.v.chadwick@kent.ac.uk

Abstract. Nanocrystalline samples of $Ba_{1-x}Ca_xF_2$ prepared by high-energy milling show an unusually high F^- ion conductivity, which exhibit a maximum in the magnitude and a minimum in the activation energy at $x = 0.5$. Here, we report an X-ray absorption spectroscopy (XAS) at the Ca and Sr K edges and the Ba L_3 edge and a molecular dynamics (MD) simulation study of the pure and mixed fluorides. The XAS measurements on the pure binary fluorides, CaF_2, SrF_2 and BaF_2 show that high-energy ball-milling produces very little amorphous material, in contrast to the results for ball milled oxides. XAS measurements of $Ba_{1-x}Ca_xF_2$ reveal that for $0 < x < 1$ there is considerable disorder in the local environments of the cations which is highest for $x = 0.5$. Hence the maximum in the conductivity corresponds to the composition with the maximum level of local disorder. The MD calculations also show a highly disordered structure consistent with the XAS results and similarly showing maximum disorder at $x = 0.5$.

1. Introduction

There are numerous reports in the literature that atomic migration in nanocrystals is unusually fast compared to the parent bulk. In ionic materials the origin for enhanced diffusion has been assigned to atomic disorder along interfaces and surfaces [1-4], however, in many cases, the experimental data in many cases are ambiguous. The data for the binary oxides is particularly varied and is very dependent on the method of sample preparation [5-9]. In contrast unusually high ionic conductivities have been established in nanocrystalline alkaline earth fluorides. Conductivity measurements on pressed pellets of 9 nm particles of calcium fluoride prepared by the inert gas condensation method showed values approximately four orders of magnitude higher than bulk material [10] and were interpreted in terms of the space charge model [11] developed by Maier [12-14]. The space charge model was also used to interpret the exceptionally high ionic conductivity of barium fluoride (BaF_2) [15]. This work was followed by a report by Heitjans and co-workers of high fluoride ion diffusivity in high-energy ball-milled nanocrystalline BaF_2 and CaF_2 as well as in BaF_2 - CaF_2 composites [16]. Later work [17] revealed that nanocrystalline high-energy ball-milled BaF_2 and CaF_2 had a conductivity two orders of magnitude higher than the parent compounds. In addition, the

IOP Conf. Series: Materials Science and Engineering **80** (2015) 012005 doi:10.1088/1757-899X/80/1/012005

presence of a metastable ternary $Ba_{1-x}Ca_xF_2$ compound which retained the cubic fluorite structure was detected [17]. It was shown that this compound could be prepared in nanocrystalline form for $0 \leq x \leq 1$ by high-energy ball-milling and the conductivity had a well-defined maximum at $x = 0.5$ and a minimum in the activation energy at the same composition[18]. In addition, high speed magic angle spinning (MAS) ^{19}F NMR revealed that there were five distinct F^- sites [18]. These sites are characterized by a distinct number of Ba and Ca cations in the first coordination shell: $[Ba]_n[Ca]_{4-n}$ $(0 \leq n \leq 4)$. The mixed sites with $n = 1, 2, 3$ dominate the NMR spectra at intermediate values of x. Presumably, the mixed cation sub-lattice causes the metastability of the compounds, influences both the formation energy of, for example, F^- interstitials for example, as well as their migration energy leading to the observed fast ion conduction. A molecular dynamics simulation of the system [19] rationalised the effects as due to the large size difference between the cations, with Ca^{2+} favouring the formation of nearest neighbour interstitial F^- ions and Ba^{2+} favouring the formation of neighbouring F^- vacancies.
In this contribution we report the results of X-ray absorption spectroscopy (XAS) measurements and molecular dynamics (MD) simulation studies of $Ba_{1-x}Ca_xF_2$. These probe the local structure and provide insight into the origins of the conductivity behavior.

2. Methodology

2.1 Materials
High energy ball-milling was used to prepare nanocrystalline CaF_2, SrF_2, BaF_2, $Ba_{1-x}Ca_xF_2$ (x = 0, 0.25, 0.5, 0.75 and 1), $Ca_{0.5}Sr_{0.5}F_2$ and $Ba_{0.5}Sr_{0.5}F_2$ using procedures described in reference [18]. The particles sizes as determined from the broadening of the peaks in the XRD powder patterns were typically 9-13 nm, if calculated with the Scherrer equation or 30-60 nm if the Williamson and Hall equation was used to include strain broadening. For the XAS measurements the powders were mixed with cellulose diluent and pressed into 13 mm diameter pellets.

2.2 XAS measurements
XAS scans were collected for the appropriate edge (Ca and Sr K edges, Ba L_3 edge) at room temperature on beam line B18 at the Diamond Light Source [20]. Data collection used transmission mode with ion chamber detectors. Continuous scanning (QEXAFS) was employed; an individual scan required 180 s and several scans were performed to improve the signal-to-noise ratio. The synchrotron energy and current were 3 GeV and 300 mA, respectively. The beam size at the sample was 700 × 700 microns. Powdered samples were mixed with cellulose as a diluent and pressed into 13 mm diameter pellets. The spectra were normalized in Athena and fitted to scattering models in R-space produced by FEFF in Artemis [21].

2.3 Molecular dynamics simulation
MD simulations were performed using the DL-POLY code [22]. The Born model of the ionic solid, together with full ionic charges, was used to describe the Ba-F, Ca-F and F-F interactions. The nanoparticles, which comprised 13824 anions and 6912 metal cations, were generated using simulated amorphisation and recrystallisation [23]; three model nanoparticle solid solutions were generated: $Ba_{0.25}Ca_{0.75}F_2$, $Ba_{0.5}Ca_{0.5}F_2$ and $Ba_{0.75}Ca_{0.25}F_2$.

IOP Conf. Series: Materials Science and Engineering **80** (2015) 012005 doi:10.1088/1757-899X/80/1/012005

3 Results and discussion

3.1 XAS measurements

All the spectra were analysed to yield the details of the local structure, i.e. bond lengths, Debye-Waller factors, etc., however the qualitative information is best represented by the Fourier transform (FT) of the k^3 weighted normalised EXAFS and these will be used in this short paper. Firstly, we consider the FTs of nanocrystalline and bulk CaF_2 shown in Figure 1. Very similar results were found for SrF_2 and BaF_2 (not shown here). There is very slight attenuation of the peaks in the nanocrystalline material, which is a distinct contrast to the XAS data for ball-milled oxides [5, 6, 24]. This indicates that the high-energy ball-milled binary fluorides are highly ordered with minimal amounts of amorphous material, possibly due to stronger ionic bonding in the fluorides. It seems reasonable to assume the high-energy ball-milled ternary $Ba_{1-x}Ca_xF_2$ compound has a similar low amorphous content.

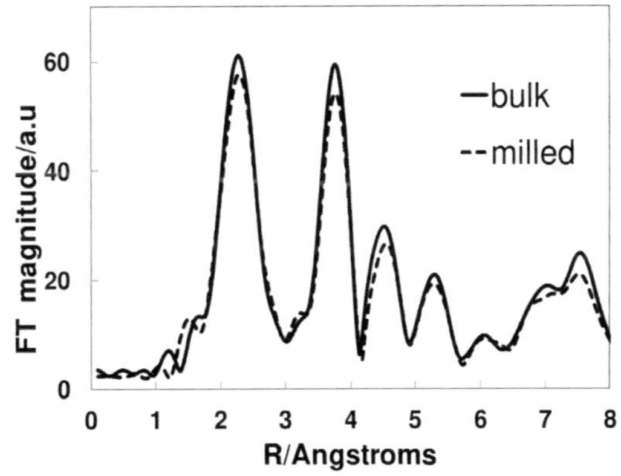

Figure 1. The FT of the normalised Ca K edge EXAFS of CaF_2.

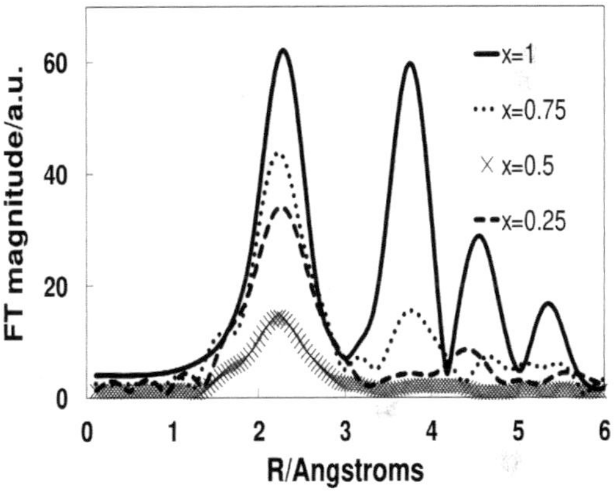

Figure 2. The FT of the normalised Ca K edge EXAFS of $Ba_{1-x}Ca_xF_2$.

The FTs of the Ca K edge EXAFS of the $Ba_{1-x}Ca_xF_2$ samples are shown in Figure 2. The two dominant peaks in the plot at ~ 2.3 Å and ~ 3.8 Å correspond to the nearest neighbor shell of eight F^- ions and the shell of next-nearest neighbour cations, respectively. These peaks are clearly attenuated for the samples with $x < 1$ as a result of the disorder created in the local structure due the replacement of Ca^{2+} by the larger Ba^{2+} cation. The effect is greatest for $x = 0.5$, where the second peak has almost disappeared, showing this sample has the highest level of disorder. Similarly the Ba L_3 edge EXAFS of these samples showed that the FT was most attenuated for the sample with $x = 0.5$. The maximum observed in the conductivity of this system is also found at $x = 0.5$ [18] and it is reasonable to associate this with the maximum level of local disorder.

3.2 MD simulations

Analysis of the model nanoparticles revealed all three $Ba_{1-x}Ca_xF_2$ samples ($x = 0.25$, $x = 0.5$ and $x = 0.75$) were crystalline and conform to the fluorite crystal structure. The particles were highly disordered with F^- anions located far from their lattice sites; domains rich in either Ca or Ba showed relatively less disorder of the local anion sub-lattice compared to domains comprising Ca and Ba on neighbouring cation sites.

The nanoparticle with $x = 0.5$ comprised the highest level of sub-lattice disorder, Figure 3, in accord with the EXAFS data. There is a higher probability of finding Ca and Ba ions on neighbouring lattice sites for the $x = 0.5$ system, compared to either the $x = 0.25$ or $x = 0.75$ system, and therefore a higher probability of a disordered anion sub-lattice. The maximum disorder in the anion sub-lattice, at $x = 0.5$, coincides with the maximum in the conductivity. Accordingly, we attribute the high ionic conductivity to the disorder of the anion sub-lattice.

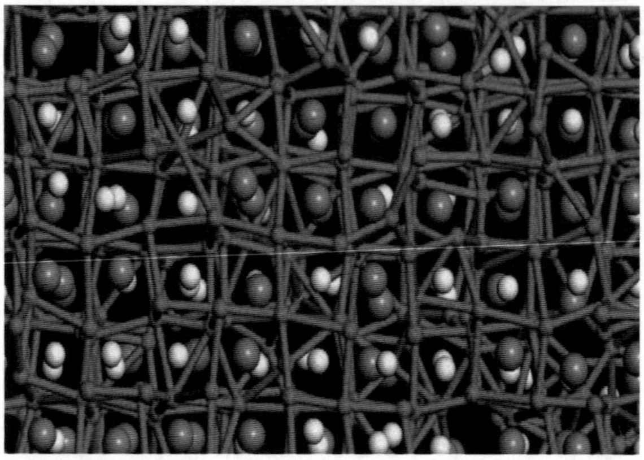

Figure 3. Ball and stick model of a segment of the $Ba_{0.5}Ca_{0.5}F_2$ nanoparticle showing the considerable disorder of the F^- anion sub-lattice. Ba^{2+} ions are coloured blue, Ca^{2+} ions are yellow and F^- ions are red. The sticks, linking the F^- ions, are included to help visualize the fluorite crystal structure.

4. Conclusions

The important outcomes of this combined XAS and MD study are as follows:-

1. High-energy ball-milled CaF_2, SrF_2 and BaF_2 contain very small amounts of amorphous material.
2. $Ba_{1-x}Ca_xF_2$ for $1 < x > 0$ is disordered with the maximum disorder at $x = 0.5$, which is the composition of the maximum ionic conductivity.

References

[1] Tuller H 2000 *Solid State Ionics* **131** 143
[2] Indris S, Heitjans P, Roman H E and Bunde A 2000 *Phys. Rev.Lett.* 84 2889
[3] Guo X, Tan C and Yuan R-Z 1995 *J. Eur. Ceram. Soc.* **15** 25
[4] Heitjans P and Indris S 2003 *J. Phys.: Condens. Matter* **15** 1257
[5] Chadwick A V and Savin S L P 2009 *Solid State Electrochemistry I*, ed. V V Kharton (Weinheim: VCH-Wiley) p 79
[6] Chadwick A V and Savin S L P 2010 *Low Dimensional Solids*, ed. D W Bruce, D O'Hare and R I Walton V V Kharton (Chichester: Weinheim: Wiley) p 1
[7] Aydin H, Korte C and Janek J 2013 *Sci. Technol. Adv. Mater.* **14** 035007
[8] Gerstl M, Frömling T, Schintlmeister A, Hutter H and Fleig J 2011 *Solid State Ionics* **184** 23
[9] Schichtel N, *et al* J 2010 *Phys. Chem. Chem. Phys.* **12** 14596
[10] Puin W and Heitjans P 1995 *Nanostruct. Mater.* **6** 885
[11] Puin W, Rodewald S, Ramlau R, Heitjans P and Maier J 2000 *Solid State Ionics* **131** 159
[12] Maier J 1987 *J. Electrochem. Soc.* **134** 1524
[13] Maier J, Prill S and Reichert B 1988 *Solid State Ionics* **28–30** 1465
[14]` Maier J 1995 *Prog. Solid State Chem.* **23** 171
[15] Sata N, Eberman K, Eber K and Maier J 2000 *Nature* **408** 946
[16] Ruprecht B, Wilkening M, Steuernagel S and Heitjans P 2008 *J. Mater. Chem.* **18** 5412
[17] Ruprecht B, Wilkening M, Feldhoff A, Steuernagel S and Heitjans P 2009 *Phys. Chem. Chem. Phys.* **11** 3071
[18] Düvel A, Ruprecht B, Heitjans P and Wilkening M 2011 |*J. Phys. Chem. C* **115** 23784
[19] Zahn D, Heitjans P Maier J *Chem. Eur. J.* 2012 **18** 6225
[20] Dent A J, *et al* 2013 Journal *of Physics: Conference Series* **430** 0120231
[21] Ravel B and Newville M 2005 *J. Synchrot. Radiat.* **12** 537
[22] Smith W and Forester T R 1996 *J. Mol. Graphics* **14** 136
[23] Sayle D C, Mangili B C, Price D W and Sayle T X 2010 *Phys. Chem. Chem. Phys.* **12** 8584
[24] Chadwick A V, *et al* 2003 *J. Phys.: Condens. Matter* **15** 431

IOP Conf. Series: Materials Science and Engineering **80** (2015) 012006 doi:10.1088/1757-899X/80/1/012006

Disorder driven transition beyond two-component, single-flavor Dirac physics in silicene

Partha Goswami

Deshbandhu College, University of Delhi, Kalkaji, New Delhi-110019, India
The corresponding author's e-mail address: physicsgoswami@gmail.com

Abstract. We consider the topological insulator (TI) phase of silicene in the presence of an applied electric field E_z perpendicular to its plane. These are assembly of Dirac fermions on a low-buckled 2D hexagonal lattice. The buckled structure generates a staggered sub-lattice potential between silicon atoms at A sites and B sites for the applied field. The starting point of our investigation is the Hamiltonian involving the Dirac kinetic energy, a mass gap term, and the spin-orbit coupling. Tuning of E_z allows for rich behaviour starting from a TI state to a band insulator(BI)with a valley-spin-polarized metal (VSPM) at a critical value in between. As long as the impurity potential strength V_0 is of the same order as the intrinsic spin-orbit coupling (SOC) t_{so}(~ 4 meV), VSPM phase is protected. The effective "two-component, single-flavor Dirac physics" remains valid in this phase.The increase in V_0, however, leads to the disappearance of this phase due to enhanced scattering processes, i.e. the out-of-plane electric field driven topological insulator to metal transition, being predicted for our system, will not be observable when the impurity potential is very high. The enhancement in SOC (t_{so}) or the Rashba spin-orbit coupling effect (t_2), however, does not lead to such a scenario. **PACS numbers:** 71.90.+q,73.43.-f,73.22.-f,71.70. Ej,85.75.-d.

1.Introduction The purpose of this communication is to report the investigation of the effect of elastic scattering by non-magnetic impurities on the transition of Dirac fermions on a low-buckled 2D hexagonal lattice, such as the stand-alone, low-buckled monolayer silicene(MLS),from a topological insulator to a semi-metal. Fleurence et al.[1] possibly have succeeded for the first time in preparing silicene on Si wafers via a conductive ceramic zirconium diboride (ZrB_2) buffer layer, though there have been quite a few independent silicene synthesis reports around the same time on the conducting substrate, such as that of the epitaxial silicene sheets on silver (111) [2,3]. It must be mentioned here that the existence of Dirac fermions itself in silicene is somewhat speculative due to the lack of direct evidences from experiment [4.5]. The MLS is basically a semimetal because the valence and conduction bands touch at the Fermi level. It consists of a honeycomb lattice of silicon atoms with two sub-lattices made of A and B sites. The states near the Fermi energy are π orbitals residing near the Dirac points K and K′ at opposite corners of the hexagonal Brillouin zone. The silicene sheet, in fact, has linear band crossing at the K and K′ symmetry points. As in graphene, thus, the charge carriers in silicene behave like relativistic particles with a conical energy spectrum and Fermi velocity $v_F \approx 10^6$ m-s^{-1}. The honeycomb lattice of the system is distorted due to a larger(than carbon) ionic radius of silicon atom and forms a buckled structure as already mentioned. The A and B sites per unit cell form two sub-lattices separated by a perpendicular distance, say, 2ℓ. The structure generates a staggered sub-lattice potential $2\ \ell E_z$ between silicon atoms at A sites and B sites for an applied electric field E_z. Silicene has a larger spin-orbit coupling (SOC)induced gap than graphene due to its buckled structure. Remarkably, the silicene exhibits a tunable band gap due to the applied electric field E_z perpendicular to the system-plane. Tuning of E_z, allows for rich behavior varying from a topological insulator(TI) to a band insulator(BI) with a valley spin-polarized metal (VSPM) at a critical value (E_c) in between. In fact, at the critical point with (E_z/E_c) = 1.00, the gap of one of the spin-split band pairs of a K point closes to give a Dirac point while the other pair is gapped. At the other K point the same spin band is gapped; other spin band, of course, is gapless. This is an out-of-plane electric field induced TI to semi-metal transition. This is also referred to as the topological phase transition (TPT). In this paper we show that, as long as the non-magnetic impurity scattering strength is moderate, i.e. the potential strength is of the same order or less than t_{so}, the VSPM phase characterized by the spin-valley locking is protected. The TPT is, however, remarkably sensitive to the impurity strength in the moderate and high doping regimes. The increase in the impurity strength leads to the effective disappearance of the VSPM phase. In fact, the strong impurity strength robs away the topological protection of the VSPM , and related two-component Dirac physics, against dissipation and fluctuations, and forces the system's entry into an unexplored domain. In this communication we do not explore *terra incognita*, we rather engage ourselves in the preliminary task of the integration of the impurity effect, via Green's function and Dyson's equation, into the single-particle spectrum which is the starting point of the investigation of the thermodynamic properties, the transport properties, and so on. The paper is organized as follows: In Sec. 2, a brief outline of the tight binding model of silicene is given and the low-energy excitation spectrum is obtained. This is followed by the presentation of the Born scattering approximation in Sec.3. The t-matrix approximation to deal with the impurity problem is also discussed in Sec.3. The renormalized single-particle excitation spectrum for the finite chemical potential is obtained. The result concerning sensitivity of TPT to the impurity potential could be found at the end of this section.

2. The Low-energy Excitation Spectrum We start with the silicene single-particle low energy Hamiltonian [6,7,8] expressed in terms of 4×4 Dirac matrices(γ^{μ}) in the Weyl basis as

$$H_{Si} = \sum_{\delta k,\alpha,\beta} c^{\dagger}_{\delta k,\alpha}\, \hbar_{\alpha\beta}(\delta k) c_{\delta k,\beta} \tag{1}$$

where

$$\hbar(\boldsymbol{\delta k}) = \left(\frac{\hbar v_F}{a}\right) \times [a\, v^x\, \boldsymbol{\delta k_x} + a\, v^y\, \boldsymbol{\delta k_y}] + \xi \left(\frac{\hbar v_F}{a}\right) \times [\,(\ell E'_z) \times (\gamma^5 \gamma^z \gamma^0) + t'_{so} \times (\gamma^5 \gamma^z \gamma^0 \gamma^5)]$$

$$-(M / \left(\frac{\hbar v_F}{a}\right))\, \gamma^5, \quad \Delta_z = E'_z = \frac{E_z}{\left(\frac{\hbar v_F}{a}\right)}, \quad \Delta_{soc} = t'_{so} = t_{so} / \left(\frac{\hbar v_F}{a}\right), \tag{2}$$

$$c_{\delta k,\beta} = (a_{\delta k\uparrow},\ b_{\delta k\uparrow},\ a_{\delta k\downarrow},\ b_{\delta k\downarrow}),\ v^x = \xi\, (\gamma^5 \gamma^0 \gamma^x + i\, t'_2\, \gamma^z),\ v^y = (\gamma^5 \gamma^0 \gamma^y - \xi\, t'_2 \gamma^5 \gamma^z). \tag{3}$$

The operators $a_{\delta k,\sigma}$ and $b_{\delta k,\sigma}$ ($\sigma = \uparrow\downarrow$), respectively, with momentum $\boldsymbol{\delta k}$ and spin σ correspond to the annihilation operators for the electronic density in the two independent sub-lattices A and B. The first term is the Dirac kinetic energy and the second is a symmetry-breaking term. It breaks the sub-lattice symmetry of the silicene's honey-comb structure and generates a gap. The stronger spin-orbit coupling(SOC) $\boldsymbol{t_{so}}$ in H_{Si} has its origin in the buckled structure of the system. The term t_2 corresponds to a Rashba spin-orbit coupling effect in *H*. Here ξ is the iso-spin index with $\xi = +1$ and $\xi = -1$ for the Dirac points $\boldsymbol{K}$ and $\boldsymbol{K'}$. The exchange field '*M*' arises due to proximity coupling to a ferro-magnet such as depositing Fe atoms to the silicene surface or depositing silicene to a ferromagnetic insulating substrate. The 2× **2** version[9]of H_{Si} ,valid when the valley mixing is absent, is given by

$$\hbar_{reduced}(\xi, s_z, \delta k) / \left(\frac{\hbar v_F}{a}\right) = [\,\xi a \sigma^x \delta k_x + a \sigma^y \delta k_y + (\,\xi s_z \Delta_{SOC} + \Delta_z)\, \sigma^z - (\mu a / v_F \hbar) \sigma^0]. \tag{4}$$

for the exchange field $M = 0$. Here, the Pauli matrices σ^x, σ^y, σ^z *and* σ^0 act in the space of the electrons' amplitudes on orbitals attributed to the independent A and B sub-lattices. In this approximation, the pseudo-spin is in the foreground; the iso-spin (described by the index $\xi = \pm 1$) and the real spin (described by an index $s_z = \pm 1$) are in the back-ground. It is, thus, possible to describe electrons by the effective two-component wave function. In the remaining part of the paper we shall confine ourselves to this two-component description. Here $\mu' = (\mu a/ v_F\hbar)$ is the dimensionless chemical potential of the fermion number. In writing this Hamiltonian we have ignored the low-buckled (silicene) structure generated intrinsic Rashba terms, as $t'_2 << \Delta_{SOC}$.

We find that the dimensionless spin-split bands close to a Dirac point are given by $E_k^{((U),(L))} = \pm [(a|\delta k|)^2+\{\xi s_z\Delta_{SOC} + \Delta_z\}^2]^{1/2} - \mu'$. Thus, the effect of the intrinsic SOC together with the electric field is to impart fermions with mass as they correspond to an effective staggered sub-lattice potential $V(\xi,s_z) = \{\Delta_{SOC} + \xi s_z \Delta_z\}$. The presence of a perpendicular electric field gives rise to spin-split gapped bands about $\boldsymbol{K}$ points at $(E_z/E_c) < 1$. The plot of silicene dispersion as a function of (k_xa) for $(k_ya) = 0$ and $(E_z/E_c) = 1$ for the valley index $\xi = +1$ is shown in Figure 1. At $(E_z/E_c) = 1$, if, the dispersion for

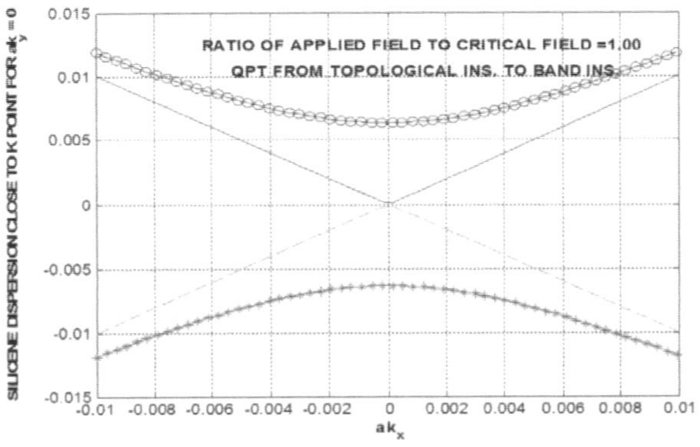

Figure 1:.The presence of spin orbit coupling and a perpendicular electric field gives rise to spin-split bands about the Dirac points, with two gaps one of which may be tuned to zero at $(E_z/E_c) = 1.00$ as shown above.

the valley (or iso-spin) index $\xi = +1$ and the spin $\downarrow$ is gapless , it is gapped for $\xi = +1$ and the spin $\uparrow$. Similarly, for $\xi = -1$ and the spin $\uparrow$ if the dispersion is gapless, for $\xi = -1$ and the spin $\downarrow$, it is gapped. The gapless situation corresponds to a valley-spin-polarized metal (VSPM) as, for $\xi = +1$ the possible spin state is $\downarrow$only, and for $\xi = -1$ the possible spin state $\uparrow$ only. For $(E_z/E_c) > 1$, the spectrum becomes fully gapped again but the system is again an insulator albeit without a gapless spectrum of edge states for the electrons protected by topology [10,11] .The transition requires the presence of the Kane-Mele [11] spin-orbit (SO) coupling t'_{so} and the electric field.

3. t-Matrix Approximation The effect of elastic scattering by non-magnetic impurities and spin-orbit scattering involve the calculation of the total self-energy $\Sigma(\boldsymbol{k},\omega_n)$ in terms of the Matsubara frequencies ω_n, which alters the single-particle excitation spectrum in a fundamental way. In the Green's function (GF) matrix $\check{G}(\boldsymbol{k}, \omega_n)$ we insert [9] the self-energy corresponding to the scattering by the former with the help of the Dyson's equation $(\check{G}(\boldsymbol{k}, \omega_n))^{-1} = (\check{G}_0(\boldsymbol{k}, \omega_n))^{-1} - \Sigma(\boldsymbol{k}, \omega_n)\, \sigma^0$, where the unit matrix σ^0 acts on the sub-lattice space. The non-interacting GF matrix may be formally given by $(\check{G}_0(\boldsymbol{k}, \omega_n))^{-1} \sim (i\omega_n\sigma^0 - \hbar_{reduced}(\boldsymbol{\xi}, \boldsymbol{s_z}, \delta k))$.We first consider only the contribution of the Fig.2(a).Assuming the elastic scattering by impurities weak,impurities are alike, and the distribution is random., we may write it as

$$\Sigma^{(1)}(\boldsymbol{k}, \omega_n)\, \sigma^0 = n_0 \sum_{k'} |V_0(\boldsymbol{k}-\boldsymbol{k'})|^2\, G_{\alpha,\alpha}(\boldsymbol{k'},\omega_n)\, \sigma^0 = \Sigma_0^{(1)}(\boldsymbol{k})\, \sigma^0 \qquad (5)$$

where $V(|\boldsymbol{q}|)$ is the Fourier coefficient of the impurity potential $V(\boldsymbol{r})$.The function $\Sigma_0^{(1)}(\boldsymbol{k})$ is the first order contribution independent of ω_n, for $\Sigma_0^{(1)}(\boldsymbol{k}) = - n_0\rho_0 (i\omega_n)\sum_{k'} |V_0(\boldsymbol{k}-\boldsymbol{k'})|^2 (\pi/|\omega_n|) = [-i\omega_n/(2|\omega_n|\tau_k)]$, where the reciprocal quasi-particle lifetime(QPLT) $\tau_k^{-1} = 2\pi n_0\rho_0\sum_{k'}|V_0(\boldsymbol{k}-\boldsymbol{k'})|^2$. The full propagator may now be written as

$$G^{(Full)}_{\alpha,\alpha}(\boldsymbol{k},\omega_n) \approx u_{\boldsymbol{k}}^2[i\omega_n - \acute{\varepsilon}_r(\boldsymbol{k}) + i(1/4\tau_k) + \mu]^{-1} + v_{\boldsymbol{k}}^2[i\omega_n + \acute{\varepsilon}_r(\boldsymbol{k}) + i(1/4\tau_k) + \mu]^{-1}, \qquad (6)$$

where the renormalized excitation spectrum is given by $\acute{\varepsilon}_r(k)=[\{(a|k|)^2+\{\Delta_{SOC}+\xi s_z\Delta_z\}^2\} - (\eta/16\,\tau_k^2)]$, $\eta=\{(a|k|)^2/(V^2+(a|k|)^2)\}$. The chemical potential μ, according to the Luttinger rule, is given by the equation $1 = \int d(\mathbf{k}a) \sum_v \rho^{(v)}{}_{Fermi}(\mathbf{k}) \times (\exp(\beta(\acute{\varepsilon}_r{}^{(v)}(\mathbf{k})-\mu))+1)^{-1}$ where $\rho^{(v)}{}_{Fermi}(\mathbf{k})$ is the Fermi energy density of states, and $\int d(\mathbf{k}a) \to {}_{-\pi}\int^{+\pi}(d(k_x a)/2\pi\ {}_{-\pi}\int^{+\pi}(d(k_y a)/2\pi$, and $\beta=(k_B T)^{-1}$.

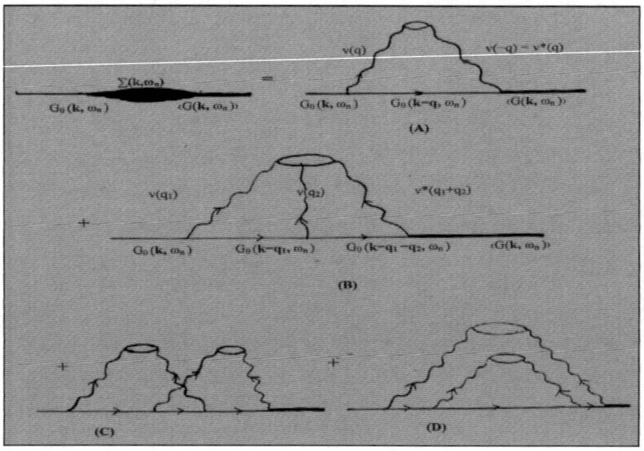

Figure 2. A few diagrams contributing to the self-energy. The wiggly lines carry momentum but no energy. The total momentum entering each impurity vertex, depicted by a slim ellipse, is zero.

Next, we consider the quasi-particle scattering problem within the t-matrix approach[12,13]. As a necessary step, assuming low concentration of impurities, one may include the contributions of all such diagrams in Fig.2 which involve only one impurity vertex. This gives the equation to determine the total self-energy $\Sigma(\mathbf{k},\omega_n)$ involving vertex function $\Gamma_0(\mathbf{k},-\mathbf{q},\omega_n)$ which in turn is given by the Lippmann-Schwinger equation:

$$\Sigma_e(\mathbf{k},\omega_n) = n_0 \sum_{k'} V_0(\mathbf{q}) G_{\alpha,\alpha}(\mathbf{k}-\mathbf{q},\omega_n)\,\Gamma_0(\mathbf{k},-\mathbf{q},\omega_n) \quad (7)$$

where the vertex function

$$\Gamma_0(\mathbf{k},-\mathbf{q},\omega_n) = V_0(-\mathbf{q}) + \sum_{q'} V_0(\mathbf{q'}-\mathbf{q})\, G_{\alpha,\alpha}(\mathbf{k}-\mathbf{q'},\omega_n)\,\Gamma_0(\mathbf{k},\mathbf{q'},\omega_n). \quad (8)$$

This is the t-martix approximation. Upon using the optical theorem for the t-matrix [13]one may write $\Sigma(\mathbf{k},\omega_n) = i\,Im\,\Gamma(k,k,\omega_n) = -i\omega_n/(2|\omega_n|\acute{\Gamma}_k)$ where $\acute{\Gamma}_k^{-1} = 2\pi\rho_0 \sum_{k'}(n_0|\Gamma_0(\mathbf{k},\mathbf{k'})|^2)$.Thus the effect of the inclusion of contribution of all the above mentioned diagrams is to replace the Born approximation for scattering by the exact scattering cross-section for a single impurity, i.e. $\tau_k^{-1} \to \acute{\Gamma}_k^{-1}$. Since $G_{\alpha,\alpha}(\mathbf{k},\omega_n)$ and $V_0(\mathbf{k})$ are known, one can determine $\acute{\Gamma}_k^{-1}$in terms of $V_0(\mathbf{k})$.

3.1 Result The increase in the non-magnetic impurity scattering strength leads to the disappearence of the VSPM phase**:** Upon considering the case $(s_z = -1, \xi = +1)$,for example, one finds the gap closing at the ***K*** point, i.e.$E_{renorm}(\mathbf{k},s_z = -1,\xi = +1) \approx \pm (a|\mathbf{k}|)$ for $\mu' = 0$, the critical Δ_z given by $\Delta_{zc} \approx [(t'_{so}{}^2 + (at'_2|\delta\mathbf{k}|)^2)^{1/2} - (\pm 1/4\,\acute{\Gamma}_k)\,\{(a|\mathbf{k}|)^2/(V^2+(a|\mathbf{k}|)^2)\}] \approx -(\pm 1/4\,\acute{\Gamma}_k)$, and the gapped spectrum at the ***K'*** point, i.e. $E_{renorm}(\mathbf{k}, s_z = -1, \xi = -1) \approx \pm[\{(a|\mathbf{k}|)^2 + 4\Delta_z{}^2\}]^{1/2}$.The fact that $\{(\left(\frac{\hbar v_F}{a}\right)a|k|)^2/(\left(\frac{\hbar v_F}{a}V\right)^2 + \left(\frac{\hbar v_F}{a}\right)a|k|)^2)\} \approx 1$ and $t'_{so}{}^2 + (a\,t'_2|\delta k|)^2) << (1/16\,\acute{\Gamma}_k^2)$ implies that $\Delta_{zc} \approx \mp 1/4\,\Gamma_k$. The result means the 'so-called' gap closing could be accessed only at a unreasonably high value of the applied electric field. The inescapable conclusion is TPT does not remain protected in the presence of strong non-magnetic impurities. With $t_2 << t_{so}$ we do not also expect the Rashba coupling to play a major role in the VSPM issue.

IOP Conf. Series: Materials Science and Engineering **80** (2015) 012006 doi:10.1088/1757-899X/80/1/012006

References

[1] A. Fleurence, R. Friedlein, T. Ozaki, H. Kawai, Y. Wang, Y. Yamada-Takamura, Phys. Rev. Lett. 108, 245501 (2012).
[2] P. Vogt, P. De Padova, C. Quaresima, J. Avila,E. Frantzeskakis, M. C. Asensio, A. Resta, B. Ealet, and G. Le Lay, Phys. Rev. Lett. 108, 155501 (2012).
[3] L. Chen, C. Liu, B. Feng, X. He, P. Cheng, Z. Ding, S. Meng, Y. Yao, and K. Wu, Phys. Rev. Lett. 109, 056804 (2012).
[4] C.-L. Lin, R. Arafune, K. Kawahara, M. Kanno, N. Tsukahara, E. Minamitani, Y. Kim, M. Kawai, and N. Takagi, Phys. Rev. Lett.110, 076801 (2013).
[5] S. Cahangirov, M. Audiffred, P. Tang, A. Iacomino,W. Duan, G. Merino, and A. Rubio, Phys. Rev. B88,035432 (2013).
[6] M. Ezawa, New J. Phys. 14 , 033003 (2012).
[7] M. Ezawa, Phys.Rev. Lett. 109, 055502(2012).
[8] M. Ezawa, Phys. Rev. Lett. 110, 026603 (2013).
[9] V.P. Gusynin, S.G. Sharapov, and A.A. Varlamov, Phys.Rev. B 90 155107(2014).
[10] N.D. Drummond, V. Zolyomi, and V.I. Fal'ko, Phys. Rev.B 85, 075423 (2012).
[11] C. L. Kane and E. J. Mele, Phys. Rev. Lett.95, 2268- 01 (2005).
[12] C. Bena, S. Chakravarty, J. Hu, and C. Nayak, Phys. Rev. B **69**, 134517 (2004).
[13] G.Rickayzen, Green's Functions and Condensed Matter(Academic Press Inc. Ltd., London,1980), chap.4,p.105.

Configurational analysis of uranium-doped thorium dioxide

A E Shields, S E Ruiz-Hernandez, N H de Leeuw

Department of Chemistry, University College London, 20 Gordon Street, London WC1H 0AJ, UK

Email: a.shields.11@ucl.ac.uk; n.h.deleeuw@ucl.ac.uk

Abstract While thorium dioxide is already used industrially in high temperature applications, more insight is needed about the behaviour of the material as part of a mixed-oxide (MOX) nuclear fuel, incorporating uranium. We have developed a new interatomic potential model, commensurate with a prominent existing UO_2 potential, to conduct configurational analyses of uranium-doped ThO_2 supercells. Using the GULP and Site Occupancy Disorder (SOD) computational codes, we have analysed the distribution of low concentrations of uranium in the bulk material, but have not observed the formation of uranium clusters or a single dominant configuration.

1. Introduction

Although the majority of nuclear power reactors operating today use uranium and/or plutonium-based fuels such as low-enriched uranium (LEU) or mixed oxide fuels (MOX), alternative nuclear fuels are being considered as well.[1] Thorium compounds can be used to fuel nuclear reactors and they present an attractive alternative to the standard uranium and plutonium-MOX fuels. As many countries around the world rethink their nuclear energy programs in the wake of the Fukushima event, interest in thorium energy is growing as a result.[2]

While thorium is a fertile material, it is not fissile and therefore a sustained conversion to the fissile ^{233}U isotope can only occur in the presence of a neutron source. Although plutonium and uranium may serve as the source, there are several advantages to blending thoria with urania that have been confirmed experimentally, including changes in the decay heat, melting point, and thermal conductivity of the material, in addition to a decrease in the release rate of fission gases despite an actual increase in the production of fission gases.[3]

Since a small amount of a neutron source is required for application as a nuclear fuel, it is useful to study thorium dioxide (thoria) in the context of uranium-doping. Pure thoria is a ceramic material with a melting point of 3651 K, the highest of any known oxide.[4] Unlike uranium and plutonium oxides, ThO_2 also represents the highest oxidation state of the material, which makes it exceptionally stable in the presence of oxygen or oxygenated water. This corrosion resistance has positive implications for the use of thoria in nuclear fuels.[3]

Thoria has the fluorite structure, space group $Fm\bar{3}m$ (225), which is shared with the dioxides of, for example, cerium, plutonium, and uranium. As ThO_2 and UO_2 are isostructural and blends of the two materials are combined to form nuclear fuels, there is significant interest in thoria-urania solid solutions, both experimentally and theoretically.

2. Methodology

To investigate the distribution of uranium atoms in thoria, a 96-atom 2x2x2 thoria supercell was used as the base material. The oxygen-oxygen interaction is described by the potential published by Catlow, whereas initial values of A and ρ were taken from the Catlow U-O Buckingham potential.[5] Due to the similarities between urania and thoria, these values provided a good starting point for the derivation of a Th-O potential, which is described by a Buckingham equation:

$$V = A\exp(-r/\rho) \tag{1}$$

where A = 1281.775, ρ = 0.391

We have next used this ThO_2 potential along with the leading Catlow UO_2 potential in the Site-Occupancy Disorder (SOD) code, version 0.26.[6] to obtain the distribution of uranium in a range of concentrations in the thoria lattice. The SOD code takes advantage of internal symmetry operators in the material to reduce the total number of possible configurations for a given substitution to only the number of independent configurations.

For a uranium-substituted 2x2x2 supercell of $Th_{(32-x)}U_xO_{64}$, where x ranges from 2-5, the number of these inequivalent configurations was determined with SOD, followed by energy minimizations under constant pressure using Gulp 3.4.[7,8] The resulting energies of each of the independent configurations was compared to pure thoria to determine the most energetically favourable distributions of uranium atoms in the system.

Additionally, the SOD code was used to determine the Boltzmann probabilities of the non-equivalent configurations.[6,9] The probability P_m of each independent configuration, m, is calculated according to equation 2, where G_m is the vibrational free energy, Ω_m represents the number of times the independent configuration occurs in the total configurational space, k_b is Boltzmann's constant, and Z is the partition function:

$$P_m = \frac{1}{Z}\Omega_m \exp\left(-\frac{G_m}{k_b T}\right) \tag{2}$$

3. Results

The SOD code identified five independent configurations in the $Th_{(32-x)}U_xO_{64}$ supercell where x= 2, whereas there are 14 for x = 3. These systems represent concentrations of approximately 6% and 9% uranium, respectively, which is in line with real-world fuel applications of doped thoria. For each value of x, the energy differences between the lowest and highest configurations were less than 0.1 eV. As a result, the Boltzmann probability distributions illustrated in Figure 1 indicate that there is not one dominant configuration for x= 2 or x=3. In all cases, the lack of one dominant configuration indicates that the system is disordered. However, by comparing the probabilities and the configurational energies, it is noted that those configurations with the highest probability of occurring are among the lowest energy structures for every value of x.

The key finding from these calculations is that the uranium atoms are distributed throughout the cell and do not form clusters in the lowest energy configurations. This indicates that any stabilisation effects of forming uranium clusters within the ThO_2 matrix are overcome by the likely entropic effects of the less-ordered structures where the U-U distances are maximised.

IOP Conf. Series: Materials Science and Engineering **80** (2015) 012007 doi:10.1088/1757-899X/80/1/012007

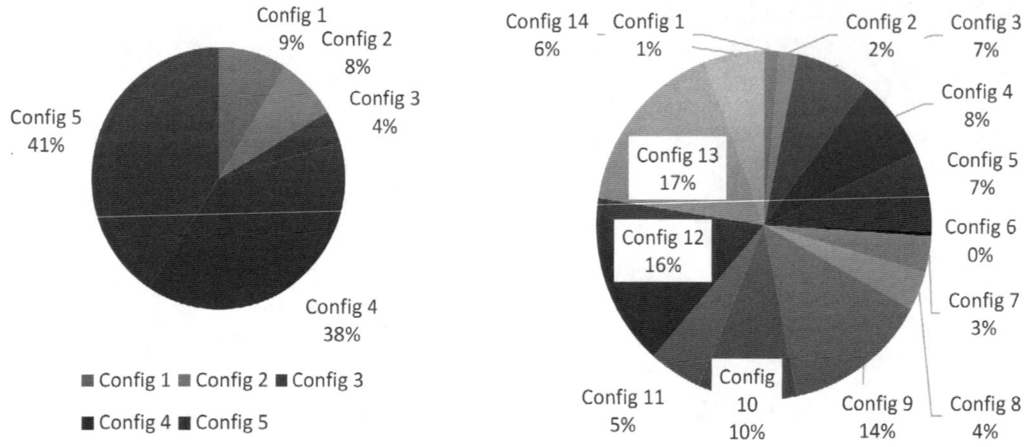

Figure 1. Probability distribution of the independent configurations of $Th_{30}U_2O_{64}$ and $Th_{29}U_3O_{64}$

4. Conclusions:

We have developed a new interatomic potential that can be successfully used in conjunction with a leading UO_2 potential to model uranium-doped ThO_2. In this study we have applied the potential model to the study the distribution of uranium atoms in thoria supercells. We have shown that uranium atoms do not cluster in the bulk material but are instead distributed throughout the cell.

Acknowledgements:

Via our membership of the UK's HPC Materials Chemistry Consortium, which is funded by EPSRC (**EP/L000202**), this work made use of the facilities of HECToR and ARCHER, the UK's national high-performance computing service, which is funded by the Office of Science and Technology through EPSRC's High End Computing Programme.

References:

[1] Crawford D C, Porter D L and Hayes S L 2007 Fuels for sodium-cooled fast reactors: US perspective *J. Nucl. Mater.* **371** 202–31

[2] Glaser A 2011 After Fukushima: Preparing for a More Uncertain Future of Nuclear Power *Electr. J.* **24** 27–35

[3] Herring J S, MacDonald P E, Weaver K D and Kullberg C 2001 Low cost, proliferation resistant, uranium–thorium dioxide fuels for light water reactors *Nucl. Eng. Des.* **203** 65–85

[4] Ronchi C and Hiernaut J P 1996 Experimental measurement of pre-melting and melting of thorium dioxide *J. Alloys Compd.* **240** 179–85

[5] Catlow C R A 1977 Point Defect and Electronic Properties of Uranium Dioxide *Proc. R. Soc. A Math. Phys. Eng. Sci.* **353** 533–61

[6] Grau-Crespo R, Hamad S, Catlow C R A and Leeuw N H de 2007 Symmetry-adapted configurational modelling of fractional site occupancy in solids *J. Phys. Condens. Matter* **19** 256201

[7] Gale J D 1997 GULP: A computer program for the symmetry-adapted simulation of solids *J. Chem. Soc. Faraday Trans.* **93** 629–37

[8] Gale J D and Rohl A L 2003 The General Utility Lattice Program (GULP) *Mol. Simul.* **29** 291–341

[9] Ruiz-Hernandez S E, Grau-Crespo R, Ruiz-Salvador A R and De Leeuw N H 2010 Thermochemistry of strontium incorporation in aragonite from atomistic simulations *Geochim. Cosmochim. Acta* **74** 1320–8

Optical Properties of Irradiated Topaz Crystals

V. Skvortsova, N. Mironova- Ulmane, L. Trinkler

Institute of Solid State Physics, University of Latvia

e-mail: vera@cfi.lu.lv

Abstract. The results of an investigation of UV-Visible absorption and photoluminescence spectra of colorless topaz before and after neutron irradiation, natural blue topaz from Ukraine, and yellow topaz are presented. We assume that the absorption band ~ 620 nm and broad emission band 300-700 nm in topaz crystals are associated with exchange interaction between a radiation defect (anion vacancies, which capture one or two electrons) and impurity ions Cr^{3+}, Fe^{3+} and Mn^{2+}.

1. Introduction

Even small amounts of impurities and structural defects have a significant effect on the physical properties of crystals, chemical and radiation resistance of products and devices manufactured on the basis of topaz crystals. Topaz crystals have been investigated as a possible candidate for dosimetry [1].Topaz is an aluminium fluorosilicate with a general composition of $Al_2(SiO_4)(F,OH)_2$. The structure of topaz consists of SiO_4 groups linking octahedral chains of $Al[O_4(F,OH)_2]$ in a zigzag fashion parallel to the crystalline c-axis. Four of six anions surrounding Al^{3+} ion belong to a SiO_4 tetrahedron and the remaining two anions are F^- or OH^-group. Topaz crystallizes in the orthorhombic system, space group Pbnm (D^{16}_{2h}) [2, 3] with four molecules per unit cell, but the OH/F substitution turns its symmetry into triclinic (P1) [4]. Coloration of topaz crystals is possible through the incorporation of transition metal impurities or by irradiation. Blue topaz now is one of the gem industry's most commercialized gemstones. The origin of the blue color in topaz is not well understood. Some authors attributed blue color to a broad absorption band generated by the so-colled R-centers (two F-vacancies with two trapped electrons [5], whereas others attributed it to an O^- center interacting with two Al ions of the topaz structure. This O^- center is produced by the irradiation in the hydroxyl sites, which substitutes for fluorine in the topaz structure [6-9]. As suggested by Priest V. at al [10] the blue defect in neutron damaged topaz is the doubly occupied dangling silicon bond.

We present a study of optical properties of topaz crystals from India and natural topaz crystals from Ukraine before and after fast neutron irradiation.

2. Materials and Methods

Yellow topaz from India and natural colorless and blue topaz from Ukraine were investigated. The samples thickness was in the range of 1.8 –3.4 mm.

The neutron irradiation was performed at the Latvian 5 MW water-water research reactor. The fluence of fast neutrons with energy > 0.1 MeV was in the range $10^{14} – 5{\cdot}10^{17}$ cm^{-2}. A cadmium filter

IOP Conf. Series: Materials Science and Engineering **80** (2015) 012008 doi:10.1088/1757-899X/80/1/012008

was used for absorption of thermal neutrons. The sample temperature did not exceed 40°C during the neutron irradiation.

The photoluminescence (PL) studies were carried out using a conventional setup for spectral measurements. The setup was equipped with a UV-laser DTL-389QT (λ =263 nm), with pulse energy 4.7 mkJ at 1kHz, or deuterium lamp LDD-400 as a source of UV light and a grating monochromator MDR-2 in the excitation channel. In the case of PL measurements the luminescence signal was analyzed by Shamrock monochromator SR-303i-B using a grating with 150 lines/mm and 500 nm blaze and detected with a CCD camera DV420A-BU2 in the accumulation mode.

The standard technique for absorption measurements has been based on a "Specord 210" (Analytikjena) double-beam spectrophotometer operating in the spectral region of 190-1100 nm. Optical measurements before and after irradiation were carried out at the temperature 300 K.

3. Results and Discussion

The absorption spectra of topaz single crystals before irradiation (curve 1), after fast neutron irradiation (curve 2), natural blue topaz from Ukraine (curve 3) and yellow topaz (curve 4) are given in Figure 1. Before fast neutron irradiation in absorption spectra of colorless topaz there was observed only one band at 230 nm. After irradiation intensity of the absorption band 230 nm increases and there appear bands with maxima at 305, 430 and 620 nm and crystals obtain blue color. In natural blue topaz there was identified an intensive band at 230 nm and less intensive bands at 430 and 620 nm. In the absorption spectrum of yellow topaz there were also observed bands in the region 230-305 nm, moreover, there was observed a weak band at 500 nm. Two lines around 230 and 305 nm belong to a single defect such as silanone (=Si=O) as it was suggested in [11] basing on similar lines observed in spodumene. The increasing of fast neutrons fluence to $10^{17}cm^{-2}$ leads to an increase of the concentration of such defects. The calculations performed in [12] using $AlCrFH_7O_{11}Si$ model molecule with unrestricted CIS method to simulate UV and visible spectra shows that the theoretical spectrum of the model is in a considerably good agreement with experimental one. Excitation spectra of Cr^{3+} investigated by Tarashchan A N at al. [13] in topaz crystal show the same lines. Albuquerque A R P L at al. suggested that band 620 nm is due to a lattice defect [14].

We assume that the band ~ 620 nm includes not only so-colled R-centers (two F-vacancies with two trapped electrons) or doubly occupied dangling silicon bands, but also the bands associated with the presence of impurities of Cr^{3+}, Fe^{3+} and Mn^{2+} ions. Band 430 nm also belongs to impurity Cr^{3+}ion.

The photoluminescence spectra of topaz crystals excited by laser at λ =263 nm are shown in Fig. 2. In the photoluminescence spectra of nonirradiated topaz crystal there are two bands at 700 and 540 nm (Fig. 2, curve 1). After fast neutron irradiation the band at 700 nm disappears and a broad band 460 nm with shoulder 380 nm is observed (Fig. 2, curve 2). The fluence rise to $5\cdot10^{17}cm^{-2}$ leads to considerable reduction of intensity of a band at 460 nm (Fig. 2, curve 3). The intensive band at 460 nm with shoulder 700 nm is observed in natural blue crystals (Fig. 2, curve 4). The photoluminescence spectra of yellow topaz contain two bands with maxima 740 and 570 nm (Fig. 2, curve 5). Fig. 3. depicts the lines obtained from decomposition of the spectra of nonirradiated and irradiated by fast neutron fluence Φ = 10^{17} cm^{-2}. Photoluminescence bands in topaz crystal before irradiation are connected with Cr^{3+} (band ~ 700 nm), Fe^{3+} (band 740 nm), Mn^{2+} (band ~ 560 nm) and peroxy-radical (band ~465 nm) [15-18]. Photoluminescence spectra of natural blue topaz crystals excited by the laser (λ = 263 nm) and the excited by lamp (λ =230 nm and λ =260 nm) and results of the curve fitting (band shapes are assumed to be Gaussian) are given in Fig. 4. When exposed with UV light, topaz samples showed a very weak luminescence (Fig. 4, curves 2 and 3). Excitation with light 230 nm leads to the appearance of a 400 nm band. The same band is observed in photoluminescence spectra of irradiated topaz after curve fitting. Excitation of natural blue topaz with light of 260 nm produces a band with maxima~ 465 nm, without a shoulder at 400 nm. Photoluminescence characteristics of variously colored topazes are caused by Cr^{3+} entering the octahedral site of the topaz structure. Taking into account that the Cr incorporation in topaz is related to the F/OH distribution [19], Cr^{3+} may be coordinated differently by O_4F_2, O_4FOH or $O_4(OH)_2$. Among three Cr^{3+} complexes, CrO_4F_2,

CrO_4FOH, and $CrO_4(OH)_2$, the second has certainly a more distorted crystal field than the first and the third.

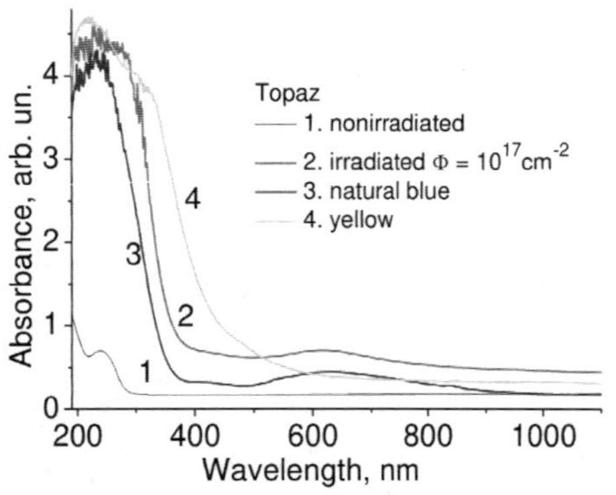

Figure 1. Absorption spectra of colorless topaz before (1) and after neutron irradiation (2), (3) natural blue topaz from Ukraine, (4) yellow topaz.

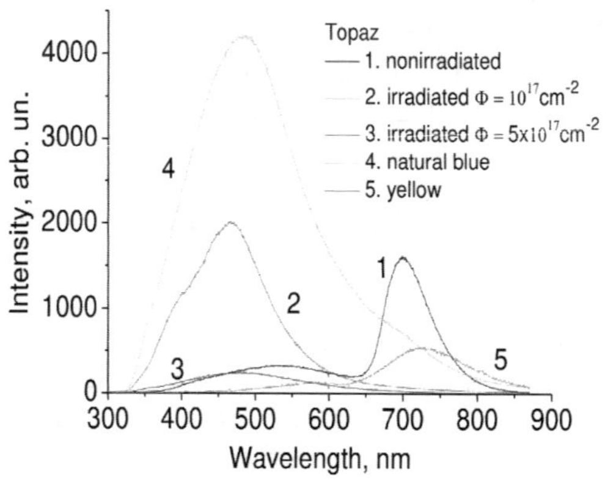

Figure 2. Photoluminescence spectra of topaz crystals: 1. before fast neutron irradiation, 2. irradiated by fast neutron $\Phi = 10^{17}$ cm^{-2}, 3. $\Phi = 5{\cdot}10^{17}$ cm^{-2}, 4. natural blue topaz 5. yellow topaz (laser excitation $\lambda = 263$ nm).

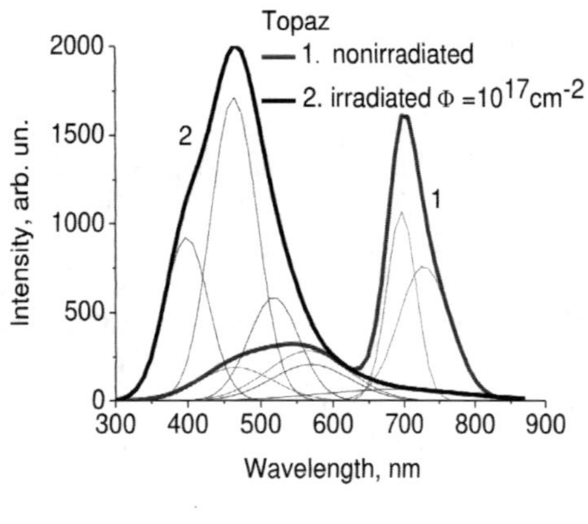

Figure 3. Photoluminescence spectra of topaz crystals: 1. before fast neutron irradiation, 2. irradiated by fast neutron $\Phi = 10^{17}$ cm^{-2} (laser excitation $\lambda = 263$ nm) and results of the curve fitting.

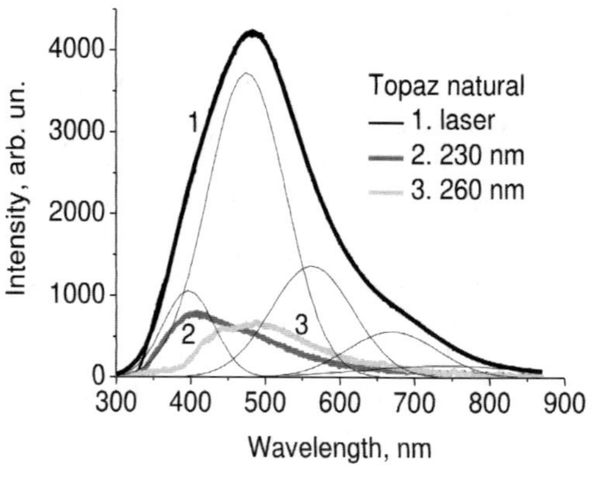

Figure 4. Photoluminescence spectra of natural blue topaz crystals: 1. after laser excitation λ = 263 nm, 2. light excitation λ =230 nm, 3. light excitation λ =260 nm and results of the curve fitting.

It is evident that the broad emission line 300-700 nm is connected with complex centers, because it appears after neutron irradiation. The presence of nearby anion vacancies, which capture one or two electrons may modify the environment at the Cr^{3+} impurity site to produce an effective increase in the crystal field. In additional to the shift in energy, the perturbation induced by the anion vacancies, which capture one or two electron (*F*- or F^+- center) can increase the oscillator strength by introducing odd components to the crystal field, or by exchange coupling between the *F*- or F^+- center and the Cr^{3+} impurity. The enhancement of the oscillator strength by the exchange interaction has been previously observed in MgF_2, $MgAl_2O_4$ and MgO [20-23]. With the concentration increase of anion vacancies, which capture one or two electrons, the number of perturbed chromium sites also increased, so that the excitation peaks became more intense as a function of neutron dose. Further fast neutron fluence increasing over $5{\cdot}10^{17}$ cm^{-2} leads to decrease of photoluminescence bands intensities - this may be connected with topaz lattice disorder [24].

4. Conclusion

The UV-Visible absorption spectra and photoluminescence spectra of yellow, natural blue and colorless topaz crystals from Ukraine before and after exposure to a fluence of fast neutrons up to $5{\cdot}10^{17}cm^{-2}$ (E>0.1 MeV) are investigated. We assume that the absorption band ~ 620 nm and broad emission band 300-700 nm in topaz crystals are associated with exchange interaction between a radiation defect (anion vacancies which capture one or two electron) and impurity ions Cr^{3+}, Fe^{3+} and Mn^{2+}.

5. Acknowledgments

This work was supported by Latvian Science Council Grant No 402/2012.

References

[1] Bomfim K S and Sousa D N 2010 *Journal of Physics: Conference Series* **249** 012057
[2] Ribbe P H and Gibbs G V 1971 *Am. Mineral.* **56** 24
[3] Northrup P A, Leinenweber R and Parise J B 1994 *Am. Mineral.* **79** 401
[4] Platonov A N, Taran M N and Balyatskii V S 1984 *Nature of color gems* (Moscow: Nedra) (in Russian).
[5] Da Silva D N, Guedes K J, Pinheiro M V B, Schweicer S, Spaeth J-M and Krambrock K 2005 *Phys. Stat. Sol. (c)* **2** 397
[6] Da Silva D N, Guedes K J, Pinheiro M V B, Spaeth J-M and Krambrock K 2005 *Phys. Chem. Minerals* **32** 436
[7] Leal A S, Krambrock K, Ribeiro L G M, Menezes M A B C, Vermaercke P and Sneyers L 2007 *Phys. Chem.Minerals* **34** 437
[8] Krambrock K, Ribeiro L G M, Pinheiro M V B, Leal A S, Menezes M A B C and Spaeth J-M 2007 *Nuclear Instruments and Methods in Physics Research A* **58** 423
[9] Pinheiro M V B, Fantini C, Krambrock K, Persiano A I C, Dantas M S S and Pimenta M A 2002 *Phys.Rew. B* ***65***, 104301
[10] Priest V, Cowan D L, Reichel D G and Ross F K 1990 *J. Applied Physics* **68** 3030
[11] Bonventi W Jr, Isotani S and Albuquerque A R P L 2012 *Advances in Condensed Matter Physics* **Article ID 873804**. doi:10.1155/2012
[12] Goto H, Niwa A, Greenhidge D C, Kato N, Ida T, Mizuno M, Endo K and Tada T 2005 *J. of Surface Analysis* **12** 249
[13] Tarashchan A N, Taran M N, Rager H and Iwanuch W 2006 *Phys. Chem.Minerals* **32** 679
[14] Albuquerque A R P L, Isotani S, Morato S P 1988 *Radiation Effects* **106** 143
[15] Taran M N Tarashchan A N, Rager H, Schott St, Schurmann K and Iwanuch W 2003 *Phys. Chem.Minerals* **30** 546
[16] Priest V, Cowan D L, Yasar H and Ross F K 1991 *Phys. Rew.B* **44** 9877
[17] Gaft M, Reisfeld R, Panczer G 2005 *Modern Luminescence Spectroscopy of Minerals and Materials* (Springer-Verlag Berlin Heidelberg)
[18] Marques C, Santos L, Falcao A N, Silva R C and Alves E 2000 *J. of Luminescence* **87-89** 583
[19] Schott S, Rager H, Schurmann K, Taran M 2003 *Eur J Mineral* **15** 701
[20] Kappers L A, Yun S I, Sibley W A 1972 *Phys. Rev. Lett.* **29** 943
[21] Shinn M D, Windscheif J C, Sardar D K Sibley W A 1982 *Phys. Rev. B* **26** 2371

[22] Cain L S, Pogatshnik, Chen Y1988 *Phys. Rev*. **37** 2645
[23] Mironova, N., Ulmanis, U. 1988 *Radiation defect and iron group metal ions in oxides* (Zinatne, Riga)(in Russian).
[24] Skvortsova V, Mironova- Ulmane N, Trinkler L and Chikvaidze G 2013 *IOP Conf. Series: Materials Science and Engineering* **49** 012051

IOP Conf. Series: Materials Science and Engineering **80** (2015) 012009 doi:10.1088/1757-899X/80/1/012009

Low temperature nanocrystalline zinc oxide for photovoltaic applications

S Greulich-Weber[1], S Arceiz Casas, and N Weber
University of Paderborn, Faculty of Science, Department of Physics, 33098 Paderborn, Germany

E-mail: greulich-weber@physik.upb.de

Abstract. Nanocrystalline ZnO was synthesized by a low temperature sol-gel route. The ZnO nano crystals were investigated by X-ray diffraction (XRD), electron paramagnetic resonance (EPR), photoluminescence (PL), and conductivity measurements. The goal was the development of nanocrystalline ZnO as a transparent conductive oxide blocking ultraviolet (uv) light and using a defect photoluminescence band for down-shifting uv light into the visible range, where organic photoactive polymers have their absorption maximum. The resulting defects and physical properties are discussed with respect to the conductivity of ZnO nanocrystals.

1. Introduction

Zinc oxide (ZnO) is a wide bandgap semiconductor with applications in photonics, optoelectronics, and sensor technology [1-5]. Doped with aluminum ZnO (AZO) attracted much attention as a promising substitute material to the currently used indium-tin-oxide (ITO) as a transparent conducting oxide (TCO). While first devices such as flat screens using AZO were recently launched, serious problems limiting the performance of devices due to unwanted defects and unintended doping still exist. Usually omnipresent defects characteristic of specific growth techniques determine optical and electrical properties of ZnO. In particular these defects compensate intended n-doping and cause strong luminescence in the visible spectral range. In most applications these implications have to be avoided. Thus, both properties were subject of detailed studies in the past [6-11].

However, there are also applications like e. g. organic photovoltaic, benefiting from such defect luminescence. Due to its wide bandgap ZnO absorbs ultraviolet (uv) light and, thus, protects photo sensitive organic materials as used in organic solar cells from being destroyed by high energy photons. At room temperature in nominally defect-free ZnO is observed strong photoluminescence of free excitons at around 3.1 eV upon uv irradiation. Regarding damageable organic materials used in photovoltaic applications this luminescence is still dangerous and at least unusable for creating a photovoltage.

Since ZnO always contains intrinsic defects a second photoluminescence band shows up upon uv irradiation [12,13]. The so-called defect luminescence band is located at around 2.2 eV. This photoluminescence band fits energetically perfect to the absorption spectrum of photoactive polymers used in organic photovoltaics. Downshifting uv light to visible light will enhance photovoltaic output. It turned out that the defect photoluminescence band can be maximized, while at the same time the unwanted free exciton luminescence is quenched. Thus creating certain defects by post-annealing or adapted crystal growth conditions allows tuning ZnO such that it acts as a uv protector and at the same time downshifting unusable uv light absorbed improving the performance of organic solar cells.

Nano and microcrystalline ZnO have been fabricated by several techniques, such as thermal decomposition (e. g. [14,15]), hydrothermal synthesis (e. g. [16,17], gas-phase reaction [18], and sol–gel method [19-21] with different particle morphologies and sizes. Electrical and optical properties as well as surface-related photocatalytic activity of ZnO nanocrystals strongly depend on dopants and defects [22]. Thus, understanding the role of intrinsic defect centers such as vacancies, interstitials, and antisites is a key toward controlling the electronic and optical properties of nanocrystalline ZnO. Electron paramagnetic resonance (EPR) has successfully been used for the identification of defects and dopants allowing the study of their microscopic and electronic structure. Together with optical properties studies of defect kinetics upon annealing are helpful for solving the described tasks.
In the present study nanocrystalline ZnO particles were prepared at low temperatures by a sol-gel procedure allowing crystal growth. A low temperature process is essential, because ZnO should act as a direct cover layer on top of the photo-active polymer, which otherwise is destroyed at temperatures above around 100°C. We report on characterization by EPR, photoluminescence (PL), conductivity measurements, and X-ray diffraction (XRD).

2. Experimental

2.1 Sample preparation

Several sol-gel routes are known for the growth of nano and micro crystalline zinc oxide [19-21]. In contrast to usual applications, AZO has to be deposited as a nanocrystalline film at rather low temperatures for organic photovoltaics. Therefore we modified a low temperature sol-gel process [44] in order to grow ZnO nanocrystals and nanocrystalline films. The low temperature sol-gel procedure [23] is based on the reaction of Zn^{2+} and OH^- in an alcoholic medium (methanol). To prepare undoped ZnO, a solutions of zinc acetate di-hydrate $Zn(ac)_2 \cdot 2H_2O$, 45 mmol, purity 99.99%) in 200 ml of methanol was prepared, aided by stirring at 52°C for 2 h. A 140 mmol KOH containing methanol solution (200 ml) was added to the solution containing $Zn(ac)_2 \cdot 2H_2O$ with constant stirring while heating at 52°C for 2h. The nanocrystals which formed were separated from the solution by centrifugation, washed several times with ethanol, and then dried in air at 25°C to obtain nanocrystalline powder. For the study of defect kinetics some nanocrystals were annealed in different atmospheres at various temperatures up to 600°C. In order to study photoluminescence and the kinetics of defects responsible for photoluminescence bands we performed annealing series including different Al doping levels. The Al doping was done during the sol-gel process by adding $Al(NO_3)_3 \cdot 9\,H_2O$.
The samples were characterized with electron paramagnetic resonance, photoluminescence, electron microscopy, X-ray diffraction, and conductivity measurements.

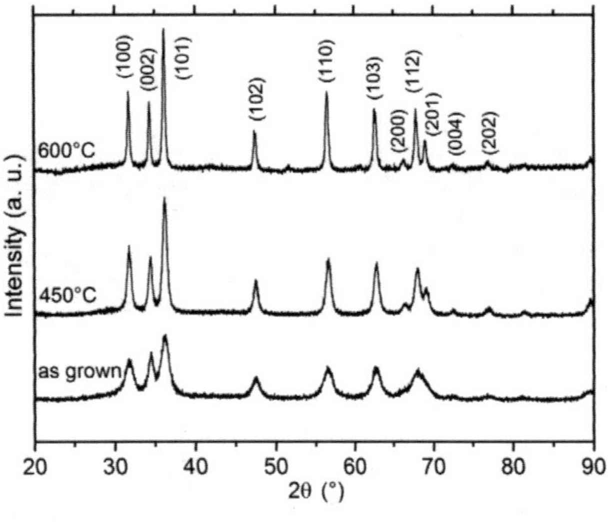

Figure 1. XRD spectra of sol-gel fabricated nanocrystalline 0.5% Al-doped ZnO. Identified signals are labelled by crystallographic direction according to [24]. The graph labelled 'as grown' was without annealing, the graphs labelled '450°C' and '600°C' are from ZnO nanocrystals annealed in argon atmosphere at the temperatures given.

2.2 Characterisation

The structure and surface morphology of sol-gel ZnO nanocrystalline powder was measured by X-ray diffraction (XRD, PANalytical X′Pert PRO with X′Celerator real-time multiple-strip detector (Cu K_α,

40 kV, 40 mA)), energy dispersive X-ray spectroscopy (EDX, Zeiss Neon 40, not shown here), and scanning electron microscopy (SEM, not shown). EDX definitely shows that the particles obtained by the sol-gel process are ZnO. The crystalline structure of the nano crystals was confirmed to be wurzite by X-ray diffraction analysis. Figure 1 shows examples from different annealing temperatures (annealing in argon atmosphere) of sol-gel ZnO nanocrystals. Already without annealing (figure 1, lower trace) XRD shows the typical reflection peaks of wurzite ZnO. These broad lines become narrower with increasing annealing temperatures (figure 1 upper traces) reflecting the growing size of the crystalline particles. Using Scherrer's equation [25] we calculated particle diameters of 8 ± 1 nm (0°C), 13 ±1 nm (450°C), and 23 ± 1 nm (600°C).

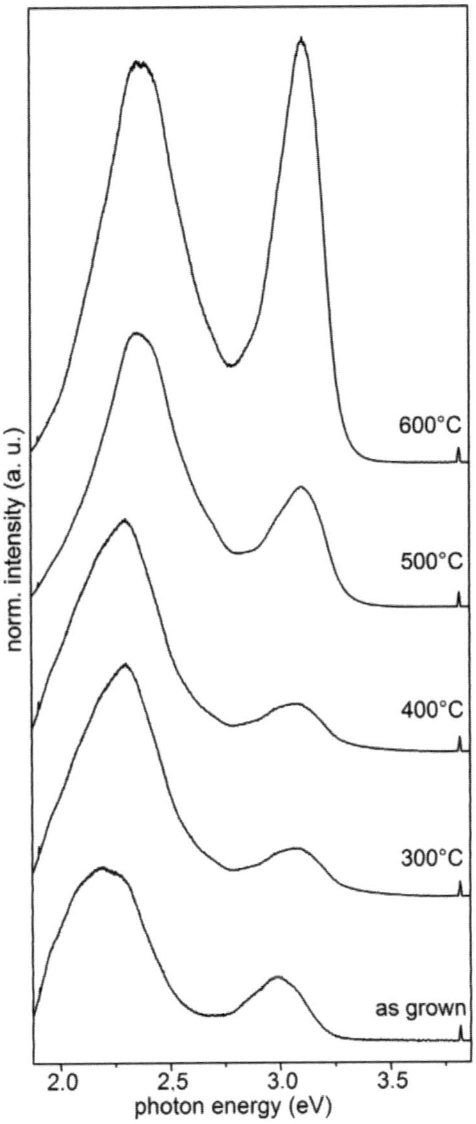

Figure 2. Photoluminescence spectra of ZnO nanocrystals measured at room temperature and prepared by a low-temperature sol-gel procedure (see text). The PL was excited with sub-bandgap light of a HeCd-LASER (see LASER peak at 3.815 eV).

Photoluminescence of the sol-gel ZnO nanocrystals was measured upon excitation with the light of a He-Cd-LASER (see figure 2). The low energy luminescence band at around 2.2 eV generally is due to defects of various origins, while the band around 3.1 eV is due to free excitons in ZnO. Both bands increase in intensity upon annealing of the ZnO nanocrystals and both are well-known from literature [22]. The exciton band has low intensity if ZnO particles are small and becomes intense upon growing particle size. A slight shift of the photoluminescence band towards higher energy is due to an increasing band gap when the size of the crystals increases. The low energy photoluminescence band (defect photoluminescence band) is also increasing in intensity upon increasing particle size; however, the band shows various small energy shifts and unresolved details within the line shape. The origin of the luminescence is still under discussion and not completely clear. Nevertheless, this band is more intensive than the exciton band in nanocrystals prepared without annealing (figure 2, lowest trace, labelled '0°C'). Upon annealing of ZnO nanocrystals in argon atmosphere from 300 to 600°C (see figure 2, further traces) the high energy exciton photoluminescence band becomes stronger, until at 600°C it shows at least the same intensity as the low energy defect luminescence band.

Further defect analysis was performed by EPR. Figure 3 shows an example for a series of nanocrystalline ZnO samples nominally doped by 0.5% Al. Already without annealing an EPR line is observed at the g-factor of shallow donors. The line shape is clearly 'Dyson' type [26], thus showing that the EPR is due to conducting electrons. A first annealing step at 100°C almost quenches the signal; while after 300°C annealing numerous new EPR lines appear. They are labelled in figure 3 according to EPR spectra of defects already known from literature (see e.g. [22]). The intense EPR line labelled 'SD' is due to shallow donors in ZnO, which are intrinsic and extrinsic having all very similar g-factors and are usually not resolved by EPR.

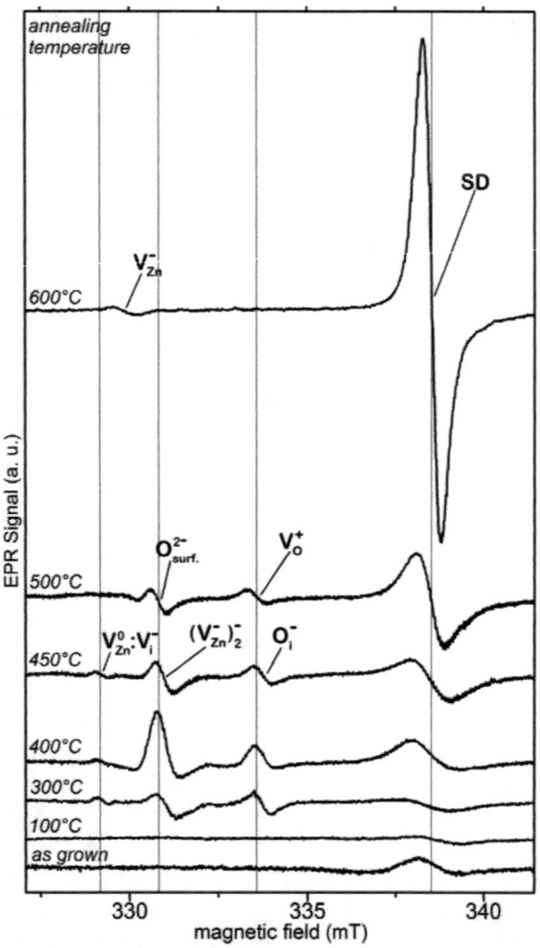

Figure 3. EPR spectra, measured at 15K, of nominally 0.5% Al doped nanocrystalline ZnO for different annealing temperatures. The EPR signal labelled 'SD' is due to shallow donors.

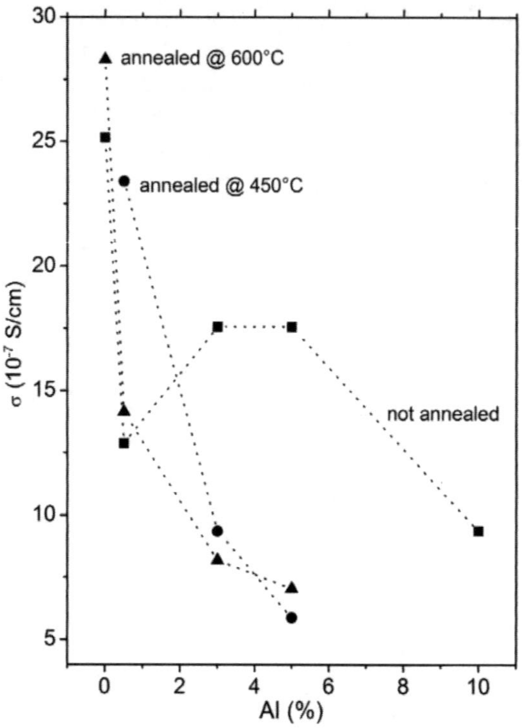

Figure 4. Conductivity of sol-gel ZnO nanocrystals measured as a powder after [22]. The samples were prepared with different Al doping levels and annealed at two different temperatures. Note that conductivity measurements of nanocrystalline ZnO films on substrates result in different conductivities, because of unpredictable current paths through grain boundaries and surface properties.

3. Discussion

Our XRD investigations (see figure 1) show that the size of ZnO nanocrystals increases upon annealing. Likewise we observed structural development of lattice defects, such as vacancies (either oxygen or zinc) or surface defects (like O^{2-}) by EPR. Thus, annealing at around 600°C results in a strong resonance of shallow donors (see figure 3; EPR is proportional to defect concentrations), mainly because of the shallow aluminum donor doping. This unambiguously shows that the crystalline properties of ZnO become dominant at that annealing temperature. This observation is also confirmed by conductivity measurements as shown in figure 4. Annealing at 600°C leads to the highest conductivity in nanocrystalline ZnO, however, in case without annealing the conductivity is almost the same (see figure 4). Since we intend applying low temperature grown nanocrystalline ZnO, having a strong defect photoluminescence band (2.2 eV in figure 2) and weak photoluminescence at 3.1 eV, a preparation of sol-gel ZnO as described above without further annealing seems to be a satisfying choice.

Acknowledgement. This work was supported by the Deutsche Forschungsgemeinschaft (TRR 142).

[1] S. Saito, M. Miyayama, K. Koumoto, J. Am. Ceram. Soc. 68, 40–43 (1985)
[2] S. J. Pearton, D. P. Norton, K. Ip, Y. W. Heo, T. Steiner, Prog. Mater. Sci. 50 (3), 293–340 (2005)
[3] B.-S. Jeon, J.S. Yoo, J. D. Lee, J. Electrochem. Soc. 143, 3923–3927 (1996)
[4] P. L. Hower, T. K. Gupta, J. Appl. Phys. 50, 4847–4855 (1979)
[5] W. G. Morris, J. Vac. Sci. Technol. 13, 926–931 (1976)
[6] S. A. M. Lima, F. A. Sigoli, M. Jafelicci and M. R. Davolos, Int. J. Inorg. Mater. **3**, 749–754 (2001).
[7] B. X. Lin, Appl. Phys. Lett. **79**, 943–945 (2001).
[8] P. S. Xu, Y. M. Sun, C. S. Shi, F. Q. Xu and H. B. Pan, Nucl. Instrum. Methods Phys. Res. B **199**, 286–290 (2003).
[9] F. A. Selim, M. H. Weber, D. Solodovnikov, and K. G. Lynn, Phys. Rev. Lett. **99**, 085502 (2007).
[10] S. B. Orlinskii, J. Schmidt, P. G. Baranov, D. M. Hofmann, Cd. M. Doneg, and A. Meijerink, Phys. Rev. Lett. **92**, 047603 (2004).
[11] S. B. Orlinskii, J. Schmidt, E. J. J. Groenen, P. G. Baranov, C. de Mello Donega, and A. Meijerink, Phys. Rev. Lett. **94**, 097602 (2005).
[12] Y. Hu and H. J. Chen, J. Nanoparticle Res. **10**, 401 (2008).
[13] P. Sagar, P. K. Shishodia, R. M. Mehra, H. Okada, A. Wakahara, A. Yoshida, Journal of Luminescence 126 800–806 (2007).
[14] J. Antony, X. B. Chen, J. Morrison, L. Bergman, Y. Qiang, Appl. Phys. Lett. 87 241917 (2005).
[15] S. Maensiri, P. Laokul, V. Promarak, J. Cryst. Growth 289 (1) 102–106 (2006).
[16] C. Wu, X. Qiao, J. Chen, H. Wang, F. Tan, S. Li, Materials Letters 60 1828–1832 (2006).
[17] K. H. Tam, C. K. Cheung, Y. H. Leung, A. B. Djurišić, C. C. Ling, C. D. Beling, S. Fung, W. M. Kwok, W. K. Chan, D. L. Phillips, L. Ding, and W. K. Ge, J. Phys. Chem. B110 20865–20871 (2006).
[18] W. J. Li, E. W. Shi, W. Z. Zhong, Z. W. Yin , J. Cryst. Growth 203 186–196 (1999).
[19] H. Nian, S. H. Hahn, Kee-Kahb Koo, J. S. Kim, S. Kim, E. W. Shin, E. J. Kim, Materials Letters 64 157–160 (2010)
[20] O. D. Jayakumar, H. G. Salunke, R. M. Kadam, M. Mohapatra, G. Yaswant and S. K. Kulshreshtha, Nanotechnology **17** 1278–1285 (2006).
[21] A. Kołodziejczak-Radzimska, T. Jesionowski, A. Krysztafkiewicz, Physicochem. Probl. Miner. Process. 44 93-102 (2010).
[22] S. Greulich-Weber, S. J. Azam, Ch. Brodehl, and T. Baumgarten, Phys. Status Solidi 10, No. 2, 221–226 (2013)
[23] O. D. Jayakumar, H. G. Salunke, R. M. Kadam, M. Mohapatra, G. Yaswant, and S. K. Kulshreshtha, Nanotechnology **17** 1278–1285 (2006).
[24] L. N. Demyanets, L. E. Li, A. S. Lavrikov, and S. V. Nikitin, ISSN 1063_7745, Crystallography Reports, 2010, Vol. 55, No. 1, pp. 142–148.
[25] Klug HP, Alexander LE (1974), Wiley, NewYork
[26] A. I. Veinger, A. Zabrodskii, T. Tisnek, S. Goloshchapov und P. Semenikhin, Technical Physics 58, No. 12, 1806 (2013).

IOP Conf. Series: Materials Science and Engineering **80** (2015) 012010 doi:10.1088/1757-899X/80/1/012010

Modelling the concentration dependence of doping in optical materials

R A Jackson[1] **, M E G Valerio**[2]

1. School of Physical and Geographical Sciences, Keele University, Keele, Staffordshire ST5 5BG, UK

2. Department of Physics, Federal University of Sergipe, 49.100-000 São Cristóvão, Brazil

E-mail: r.a.jackson@keele.ac.uk

Abstract. As well as understanding the location of dopants in optical materials, it is also important to understand how much dopant can be added to a given material. A method for calculating the maximum concentration of dopants has been developed, and applied to dopants in mixed metal fluorides for optical and nuclear clock applications. Applications to rare earth doping in $YLiF_4$, and Th doping in $LiCaAlF_6$/$LiSrAlF_6$ are described, and compared with available experimental data.

1. Introduction

This paper is part of an ongoing modelling study of dopants in optical materials. Previous papers have addressed the question of where dopants are located in the host lattice (see e.g. [1]) and their effect on morphology [2], and an initial study of concentration dependence of dopants has been published [3].

The motivation for this study is to develop a method by which the maximum concentration of a given dopant in a host lattice can be predicted. This is useful information for device development and can potentially save time in crystal growth experiments.

The paper starts with a description of the method to be used, using rare earth doping in $YLiF_4$ as an example, and it is then applied to the doping of $LiCaAlF_6$/$LiSrAlF_6$ and CaF_2 with ^{229}Th ions, which is relevant to the development of nuclear clocks using these materials discussed previously [4, 5].

2. Methodology and Results

2.1 The doping process for rare earth ions in $YLiF_4$

[1] To whom any correspondence should be addressed.

Doping $YLiF_4$ with rare earth (M^{3+}) ions can be represented as follows (where doping at the Y site is assumed), and x is the mole fraction of dopant:

$$(1\text{-}x)\ YF_3 + x\ MF_3 + LiF \rightarrow Y_{1-x}M_xLiF_4$$

The energy of this reaction corresponds to the solution energy of the doping process, E_{sol}:

$$E_{sol} = E\ (Y_{1-x}M_xLiF_4) - [(1\text{-}x)\ E_{latt}\ (YF_3) + x\ E_{latt}\ (MF_3) + E_{latt}\ (LiF)]$$

The terms in square brackets are calculated lattice energies, E_{latt}, and the only issue is in calculating the first term in the equation. This has been done using the equation below, which effectively splits the defective lattice energy into a perfect lattice component E_p (1) and a defect formation energy term E_D^{ML}:

$$E_D(x) = x\ E_D^{ML} + E_p\ (1)$$

The final solution energy expression then becomes:

$$E_{sol} = E\ (x\ E_D^{ML} + E_{latt}\ (YLiF_4)) - [(1\text{-}x)\ E_{latt}\ (YF_3) + x\ E_{latt}\ (MF_3) + E_{latt}\ (LiF)]$$

Here the term E_D^{ML} is the substitution energy of an M^{3+} cation at the Y^{3+} site.

Using this expression, maximum dopant concentrations in mol % have been calculated for rare earth ions in $YLiF_4$, as given in table 1. These correspond to the value of 'x' when $E_{sol} = 0$.

Table 1. Calculated maximum dopant concentrations for rare earth ions in $YLIF_4$ [6]

Ion	Max. mol %	Ion	Max. mol %
La	0.69	**Tb**	1.41
Ce	0.76	**Dy**	1.28
Pr	0.85	**Ho**	1.40
Nd	0.93	**Er**	1.52
Sm	1.23	**Tm**	1.33
Eu	1.15	**Yb**	1.51
Gd	1.22	**Lu**	1.49

2.2 Th^{4+} doping in $LiCaAlF_6$

In a previous paper [4] the doping of $LiCaAlF_6$ with Th^{4+} has been discussed in the context of developing a 'nuclear clock' material which exploits the unusually low first nuclear excited state energy of the ^{229}Th isotope. It is necessary to embed the Th in a transparent crystal, and the calculations predict that Th^{4+} will substitute at the Ca^{2+} site with charge compensation by the formation of two F^- interstitials. The doping process is therefore assumed to follow this reaction:

$$x\ ThF_4 + (1\text{-}x)\ CaF_2 + LiF + AlF_3 \rightarrow LiTh_xCa_{1-x}AlF_{6+2x}$$

Using the approach described in 2.1, this leads to a solution energy expression as follows:

$$E_{sol} = E\ (LiTh_xCa_{1-x}AlF_{6+2x}) - [x\ E_{latt}\ (ThF_4) + (1\text{-}x)\ E_{latt}\ (CaF_2) + E_{latt}\ (LiF) + E_{latt}\ (AlF_3)]$$

IOP Conf. Series: Materials Science and Engineering **80** (2015) 012010 doi:10.1088/1757-899X/80/1/012010

Which leads to:

$E_{sol} = E\ (x\ E_D^{ML} + E_{latt}\ (LiCaAlF_6)) - [x\ E_{latt}\ (ThF_4) + (1\text{-}x)\ E_{latt}\ (CaF_2) + E_{latt}\ (LiF) + E_{latt}\ (AlF_3)]$

Now the term E_D^{ML} is the energy of substitution of the Th^{4+} ion at the Ca^{2+} site, plus the energy of formation of two F^- interstitials.

Setting $E_{sol} = 0$ a value of x =0. 06 or 6 mol% is obtained for the maximum dopant concentration of Th^{4+} ions in $LiCaAlF_6$. Experiments are in progress to test these conclusions.

2.3 Th^{4+} doping in CaF_2

Th^{4+} doping in CaF_2 was considered in [5], where it was again found that the predicted location of the dopant Th^{4+} would be at the Ca^{2+} site, with charge compensation by F- interstitials. The doping scheme is assumed to be:

$x\ ThF_4 + (1\text{-}x)\ CaF_2 \rightarrow Ca_{1-x}Th_xF_{2+2x}$

The corresponding expression for the solution energy is:

$E_{sol} = E\ (Ca_{1-x}Th_xF_{2+2x}) - [x\ E_{latt}\ (ThF_4) + (1\text{-}x)\ E_{latt}\ (CaF_2)]$

This can be written as:

$E_{sol} = E\ (x\ E_D^{ML} + E_{latt}\ (CaF_2)) - [x\ E_{latt}\ (ThF_4) + (1\text{-}x)\ E_{latt}\ (CaF_2)]$

For binary systems like CaF_2 a minimum value of 'x' is not calculated, but it is possible to calculate E_{sol} as a function of x. Preliminary experimental measurements [7] suggest dopant concentrations of between 0.4-0.7 mol%, corresponding to solution energies in the range 2.3 – 3.5 eV. These values are of the same order of magnitude as solution energies calculated for many doping processes in optical materials. However, more detailed measurements are in progress, and a detailed comparison will be reported in a later publication.

3. Conclusions

This paper has presented a method for predicting the maximum concentration of dopants in optical materials, and has been applied to two representative systems. It is expected that the method will be further refined once more experimental data is available, but in its present form reliable predictions have been made.

4. References

[1] Maddock E M, Jackson R A and Valerio M E G 2010 *J. Phys.: Conference Series* **249** 012040
[2] Littleford T E, Jackson R A and Read M S D 2013 *Phys. Stat. Sol. C* **10** (2) 156
[3] Rezende M V S, Valerio M E G and Jackson R A 2011 *Optical Materials* **34** 109
[4] Jackson R A, Amaral J B, Valerio M E G, DeMille DP and Hudson E R 2009 *J. Phys.: Cond. Matt.* **21** 325403
[5] Dessovic P, Mohn P, Jackson R A, Winkler G, Schreitl M, Kazakov G and Schumm T 2014 *J. Phys.: Cond. Matt.* **26** 105402

[6] Littleford T E 2014 PhD thesis, University of Keele, UK
[7] Schumm T private communication

Acknowledgements

MEGV is grateful to CNPq (Conselho Nacional de Desenvolvimento Científico e Tecnológico) for research support.

Appendix

Table of lattice energies and defect formation energies used in sections 2.2 and 2.3

Energy term	Energy/eV
E_{latt} (CaF_2)	-26.99
E_{latt} (LiF)	-10.93
E_{latt} (AlF_3)	-63.99
E_{latt} (ThF_4)	-79.98
E ($Th_{Ca}^{\bullet\bullet}$) (CaF_2)	-43.4
E (F_i') (CaF_2)	-1.9
E ($Th_{Ca}^{\bullet\bullet}$) ($LiCaAlF_6$)	-41.72
E (F_i') ($LiCaAlF_6$)	-2.01

IOP Conf. Series: Materials Science and Engineering **80** (2015) 012011 doi:10.1088/1757-899X/80/1/012011

β-ray irradiation effects on silica nanoparticles

A Alessi[1,3], S Agnello[1], G Buscarino[1], B Boizot[2], M Cannas[1], F M Gelardi[1]
[1] Department of Physics and Chemistry, University of Palermo, Via Archirafi 36, I-90123 Palermo, Italy
[2] Laboratoire des Solides Irradiés, UMR 7642 CEA-CNRS-Ecole Polytechnique, Ecole Polytechnique, 91128 Palaiseau cedex, France

e-mail: antonino.alessi@unipa.it

Abstract. By electron paramagnetic resonance (EPR) measurements, we examine the amplitude of the signal typically due to a combination of NBOHC (Non Bridging Hole Center) and POR (Peroxy Radical) defects induced by β-ray irradiation (from 1.2 to 1200 MGy) in silica nanoparticles with diameter ranging from 7 to 20 nm. Our data indicate that the signal line-shapes recorded at different doses is quite independent from the particles sizes and from the dose. Furthermore, for each considered nanoparticles size, the concentration of defects is also almost constant with respect to dose, and it does not change significantly if measured after 2 or 9 months from the irradiation. By contrast, we observed that the concentration of NBOHC+POR decreases on increasing the specific surface, indicating that the content of the defects depends on the nanoparticles size. Such dependence can be explained by a shell model in which the detected defects are located in the inner part of the nanoparticles.

1. Introduction

Radiation effects in silica have been investigated by several research groups for different reasons and from different points of view [1-7]. A relevant part of these efforts has been devoted to the understanding of the point defects properties and of their generation mechanisms [1, 8-15]. Among the most important point defects of silica we remind the E′Si center constituted by an unpaired electron localized on a sp^3 orbital of a three coordinated Si atom ($\equiv Si^\bullet$ where ≡ stands for the three single bonds with the O atoms, and $^\bullet$ represents the unpaired electron) [1], the Non Bridging Oxygen Hole Center (NBOHC) constituted by an unpaired electron localized on a 2p orbital of a non-bridging oxygen ($Si\text{-}O^\bullet$) [1, 8], and the Peroxy Radical (POR) constituted by an unpaired electron located on a pair of oxygen atoms of the structure $Si\text{-}O\text{-}O^\bullet$ [1, 8, 14]. Because of their paramagnetic nature, all of these defects have been studied by electron paramagnetic resonance (EPR) measurements. Previous investigations characterized the EPR signals of such defects [1, 8, 10] also evidencing that they can be generated by irradiation through different mechanisms involving specific precursors [1, 8, 12-14] or Si-O-Si linkages [1-3]. Furthermore, it was shown that they can be converted into other defects by the reaction with diffusing atoms or molecules [1, 8, 12, 14, 15].
As for other materials, the research on silica has been recently extended also to nanostructures. In this context the study of point defects induced by irradiation has been concerned and the investigation of the E′Si through their EPR signals and other experimental data have evidenced that silica nanoparticles

[3] Antonino Alessi, e-mail: antonino.alessi@unipa.it.

IOP Conf. Series: Materials Science and Engineering **80** (2015) 012011 doi:10.1088/1757-899X/80/1/012011

can be described by a shell model (surface shell + core) [16-23]. Apropos, until now few information are available for high irradiation doses and for the NBOHC and the POR defects in nanoparticles. For these reasons the present study is focused on the EPR investigation of the NBOHC and POR defects properties and generation in the dose range 1.2-1200 MGy in silica nanoparticles of different average diameters.

2. Experimental

We studied aerosil silica nanoparticles produced by Evonik industries, the particles are produced by oxidation of $SiCL_4$ in O_2/H_2 flame [24, 25]. The specific surface (S) and the average diameter (AD) of each type of nanoparticles have been estimated by the producer by performing BET and TEM analysis. The as-received powders have been pressed in a uniaxial press using a pressure of about 0.3 GPa. In Table 1 we report the name, the specific surface and the average diameter of each investigated type of nanoparticles.

Table 1. Name of the samples, average diameter and specific surface [24,25].

Sample name	Average Diameter (nm)	Specific Surface (m^2g^{-1})
AE90	20	90 ± 15
AE150	14	150 ± 15
AE200	12	200 ± 25
AE300	7	300 ± 30
AE380	7	380 ± 30

The samples were irradiated with β-rays using a Pelletron accelerator (electron energy 2.5 MeV) at SIRIUS irradiation facility (Ecole Polytechnique, Palaiseau, France). The irradiations were performed at a dose rate of about 18 MGy/h, the minimum dose was of about 1.2 MGy, whereas the maximum one was of 1200 MGy. All the irradiations have been performed in the temperature range 60÷70 °C. We irradiated different samples for each dose. We recorded EPR spectra by standard first harmonic mode with a CW-Bruker EMX-Micro Bay spectrometer, using a microwave frequency of about 9.8 GHz and a 100 kHz magnetic field modulation frequency. The measurements were acquired two and nine months after the irradiation at room temperature.

3. Results and Discussion

No EPR signals are detected in all the investigated samples before the irradiation [20]. In figure 1a, we report the EPR spectrum recorded for the sample AE200 irradiated up to 12 MGy. This measurement evidences the presence of the E′Si at about 3450 Gauss and of the signal of the NBOHC and/or the POR (NBOHC+POR in the following) defects. In the inset of figure 1a, we report a zoom of the most intense part of the signal of the NBOHC+POR, the arrow highlights the amplitude of the signal (Amp) used for the following analysis. To study the line-shape of the EPR signals present in the different samples irradiated at different doses we normalized each spectrum by Amp. A comparison is reported in figure 1b for the sample AE200 irradiated at different doses, and in figure 1c for the spectra recorded at the dose of 171 MGy in the different materials. These data show that the line-shape of the EPR signal is almost independent from the dose and from the sample type. Basing on these results we can suggest that the relative concentrations of NBOHC and POR defects are independent from the sample and from the dose. This finding enables to compare the overall content of NBOHC+POR as a function of the dose and of the specific surface of the nanoparticles in various samples using the values of Amp.

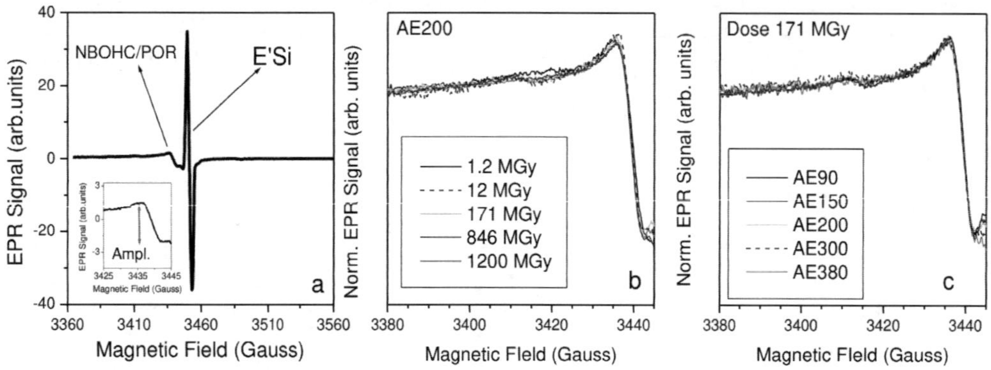

Figure 1. a) EPR spectrum recorded after two months from the irradiation at the dose of 12 MGy for the AE200 nanoparticles; **b)** Normalized EPR signals in the range 3380-3445 Gauss for the AE200 material irradiated at the doses of 1.2 (—), 12 (---), 171 (—), 846 (—) and 1200 MGy (—); **c)** spectra recorded at the dose of 171 MGy in the AE90 (—), AE150 (—), AE200 (—), AE300 (---) and AE380 (—) materials.

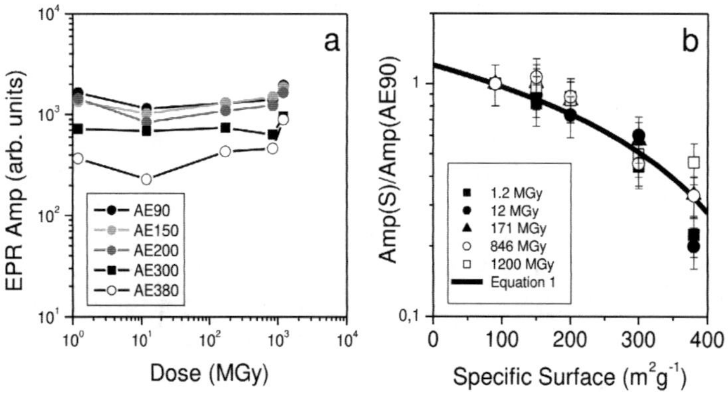

Figure2. a) Normalized EPR amplitude recorded for the samples AE90 (-●-), AE150 (-●-), AE200 (-●-), AE300 (-■-) and AE380 (-○-) as a function of the dose; **b)** ratio Amp (S)/Amp(AE90) at the doses of 1.2 (■), 12 (●), 171 (▲), 846 (○) and 1200 MGy (□), the black line indicates the values calculated by equation 1.

In figure 2a, we report the values of Amp normalized for the experimental conditions and for the samples masses as a function of the dose. The data indicate that for each material the value of Amp is almost independent from the dose suggesting that a saturation concentration value has been attained already at the lowest applied dose. By contrast, we note that the amplitude of the signal depends on the nanoparticles specific surface. More in details, as illustrated by figure 2b, the parameter Amp decreases, independently on dose, with the increase of the specific surface. To compare these results with the previous proposed core shell model [20-23] we evaluated the ratio (Amp(S)/Amp(AE90)) between the Amp measured in the various samples of different specific surface S, and the one measured in the AE90 nanoparticles irradiated at the same dose. Basing on the shell model it is expected that [20, 23]:

$$\frac{Amp(S)}{Amp(AE90)} \approx \frac{Amp^{core}}{Amp(AE90)}\left(1-\rho_S \times \delta \times S\right) \qquad (1)$$

considering that in general the total EPR signal is given by the sum of the signal arising from the defects located in the core (Amp^{core}) and in the surface shell ($Amp^{surface}$) of the nanoparticle; furthermore, in the model the ratio $\rho_s V_{ss}/m_T$ (V_{ss} being the surface shell volume, ρ_s its density and m_T the total particle's mass) is approximated by the quantity $\rho_s \cdot \delta \cdot S$ (δ being the surface shell thickness and S the specific surface) [20-23]. The line in figure 2b plots the values estimated by eq.(1) inserting $Amp^{core}/Amp(AE90) = 1.2$, $\rho_S = 2.43$ g cm^{-3} (surface shell density [16, 20-23]) and $\delta = 0.8$ nm, which was chosen after fitting the data at each dose. This latter value is slightly lower than the one (about 1nm) previously obtained [20-23], but can be considered in good agreement considering the error bars of the data. In addition, to fit the experimental data, here we assume that the number of NBOHC and POR which are induced in the surface shell are negligible.

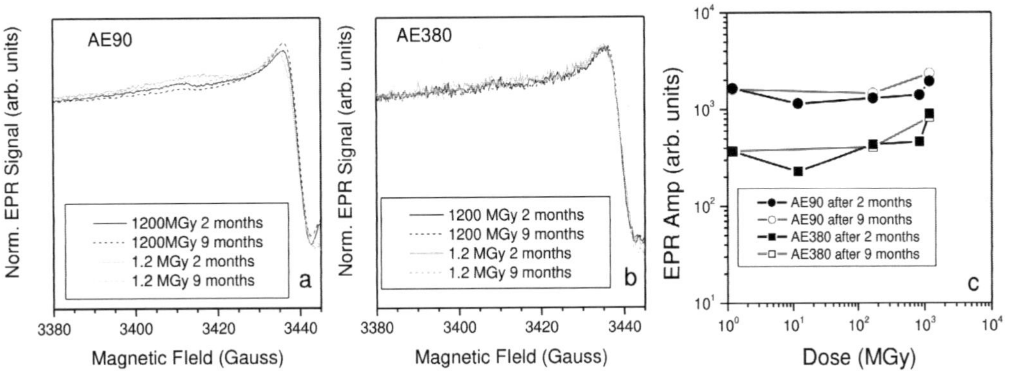

Figure 3. Normalized EPR signals in the range 3380-3445 Gauss data recorded for AE90 (a) and AE380 (b). In both panels (—) spectra acquired 2 months and (---) 9 months after irradiation at 1200 MGy, spectra acquired 2 months (—) and 9 months (---) after irradiation at 1.2 MGy; c) Amplitude of the NBOHC/POR signal as a function of the dose in the sample AE90 after 2 months (-•-) and 9 months (-○-) from the irradiation and in the sample AE380 after 2 months (-■-) and 9 months (-□-) from the irradiation.

For the samples irradiated at 1.2, 171 and 1200 MGy we repeated the measurements about nine months after the irradiation. From the comparison of the spectra reported in figure 3a and b we can guess that the spectra recorded nine months after irradiation show similar lineshape to the ones recorded two months after irradiation. In figure 3c, we compare the signal amplitude recorded for the samples AE90 and AE380 nine and two months after irradiation as a function of the dose. We note that the data are compatible within the experimental error and that for all the investigated materials the larger difference is of about 20% of the amplitude. Basing on this finding, we can consider the defects concentration unchanged.

4. Conclusion

We studied the EPR signals induced in silica nanoparticles by β irradiation up to 1200 MGy, focusing the attention on the signals attributed to NBOHC and/or POR defects. We observed that the signal recorded 2 months after irradiation does not significantly depend on the dose, it does not change significantly if measured 2 or 9 months after the irradiation and that the lineshapes of the signal recorded in different nanoparticles are comparable. By contrast, the signal amplitude depends on the specific surface of the nanoparticles. This latter result is interpreted within the shell model, which describes the nanoparticles as constituted by an inner part were the defects are mainly located and by a surface shell in which the number of defects is negligible.

Acknowledgments
Authors acknowledge all the members of the lamp group (http://www.fisica.unipa.it/amorphous/) for useful discussions, G. Napoli and G. Tricomi for technical assistance. Partial financial support by the FAE-PO FESR SICILIA 2007/2013 4.1.1.1 and by the University of Palermo FFR 2012/2013 projects is acknowledged. Irradiations were supported by EMIR accelerator network, proposal n°11-10-9841.

References

[1] Griscom D L, Friebele E J 1986 *Phys. Rev. B* **34** 7524
[2] Arai K, Imai H, Isoya J, Hosono H, Abe Y *Phys Rev B* 1992 **45** 10818
[3] Alessi A, Girard S, Cannas M, Agnello S, Boukenter A, Ouerdane Y 2012 *J. Lightw. Technol.* **30** 1726
[4] Brichard B, Borgermans P, Fernandez Fernandez A, Lammens K, Decréton M 2001 *IEEE Trans. Nucl. Sci.***48** 2069
[5] Takahashi M, Uchino T, Yoko T 2002 *J. Am. Ceram. Soc.* **85** 1089
[6] Devine R A B, Duraud J P, Dooryhée E 2000 *Structure and Imperfections in Amorphous and Crystalline Silicon Dioxide* (New York:Wiley)
[7] Yang T, Gao Y, Huang X, Zhang Y, Toulemonde M, Xue J, Yan S, Wang Y 2011 *J Non-Cryst Solids* **357** 3245
[8] Skuja L, Kajihara K, Hirano M, Hosono H 2012 *Nucl. Instrum. Methods Phys. Res., Sect. B* **286** 159
[9] Griscom D L 2001 *Phys. Rev. B* **64** 174201
[10] Griscom D L, Mizuguchi M 1998 *J Non-Cryst Solids* **239** 66
[11] Skuja L, Hirano M, Hosono H, Kajihara K 2005 *phys. stat. sol. (c)* **2** 15
[12] Messina F, Cannas M 2006 *J. Phys.: Condens. Matter* **18** 9967
[13] Imai H, Arai K, Isoya J, Hosono H, Abe Y, Imagawa H 1993 *Phys Rev B* **48** 3116
[14] Pacchioni G, Skuja L, Griscom D L 2000 *Defects in SiO_2 and Related Dielectrics: Science and Technology* (Dordrecht:KluwerAcademic)
[15] Agnello S, Cannas M, F Messina, Nuccio L, Boizot B 2009 *J Non-Cryst Solids* **355** 1042
[16] Roder A, Kob W, Binder K 2001 *J. Chem Phys* **114** 7602
[17] A. Stesmans, K. Clémer, and V. V. Afanas'ev 2005 *Phys. Rev. B* **72** 155335
[18] A. Stesmans, K. Clémer, and V. V. Afanas'ev 2008 *J Non-Cryst Solids* **354** 233
[19] Uchino T, Aboshi A, Kohara S, Ohishi Y, Sakashita M, Aoki K, 2004 *Phys. Rev. B* **69** 155409
[20] Vaccaro G, Buscarino G, Agnello S, Sporea A, Oproiu C, Sporea D G, Gelardi F M 2012 *J. Phys. Chem. C* **116** 144
[21] Alessi A, Agnello S, Buscarino G, Gelardi F M 2013 *J Non-Cryst Solids* **362** 20
[22] Alessi A, Agnello S, Buscarino G, Gelardi F M 2013 *J. Raman Spectrosc.* **44** 810
[23] Alessi A, Iovino G, Buscarino G, Agnello S, Gelardi F M 2013 *J. Phys. Chem. C* **117** 2616
[24] Basic Characteristics of Aerosil, 4th ed.; Degussa: Frankfurt, 2001
[25] Evonik industries http://corporate.evonik.com/en/Pages/default.aspx

IOP Conf. Series: Materials Science and Engineering **80** (2015) 012012 doi:10.1088/1757-899X/80/1/012012

Inertial and interference effects in optical spectroscopy

W Karstens [1] **and D Y Smith** [2, 3]

[1] Saint Michael's College, Colchester, VT 05439, USA
[2] University of Vermont, Physics, 82 University Place, Burlington, VT 05405, USA
[3] Argonne National Laboratory, 9700 South Cass Avenue, Argonne, IL 60439, USA

E-mail: dysmith@uvm.edu

Abstract. Interference between free-space and material components of the displacement current plays a key role in determining optical properties. This is illustrated by an analogy between the Lorentz optical model and a-c circuits. Phase shifts in material-polarization currents, which are *inertial*, relative to the *non-inertial* vacuum-polarization current cause interference in the total displacement current and, hence, variation in E-M wave propagation. If the displacement-current is reversed, forward propagation is inhibited yielding the semi-metallic reflectivity exhibited by intrinsic silicon. Complete cancellation involves material currents offsetting free-space currents to form current-loops that correspond to plasmons.

1. Introduction

An analogy between the optical response of matter and currents in a-c circuits suggests a novel interpretation of optical properties as the result of interference between free-space (vacuum) and matter displacement currents. The analogy reflects the similarity of the differential equations for the response of optical materials and a-c circuits. However, the underlying physics initially appears different: The Lorentz model rests on Newton's laws; electric-circuit analysis assumes Kirchhoff's laws. Further, optics focuses on electric-dipole moments; the focus in circuits is on currents.

Yet, these approaches are basically the same. Their relationship parallels that between Newtonian and energy-based methods in mechanics [1-3], but with emphasis on different dynamical variables. Significant insights into optical properties can be had by considering displacement current as well as polarization. This emphasizes the roles of vacuum and matter currents in E-M wave propagation. As with circuit branch currents, these displacement currents interfere to determine the optical response.

Interference is pronounced near material resonances and may lead to large excursions in total-current phase relative to an applied field. In the extreme, displacement-current reversal yields reflection not transmission. This may occur even in the absence of free electrons, as in some intrinsic semiconductors, e.g., silicon. Moreover, the analogy provides a picture of plasma oscillations in terms of the complete destructive interference of vacuum and matter currents. The analogy emphasizes the role of inertia in optics and optical sum rules, e.g., the *f*-sum rule [4, 5], and their circuit analogues [6].

2. Optical properties and circuits

Optical properties are commonly discussed in terms of electric susceptibilities. In the dielectric-continuum model a material's susceptibility is the sum of the susceptibilities of free space and of a uniform distribution of the matter present, $\chi(\omega)$ [7]. The free-space component is assumed to be unperturbed by the matter. The matter's susceptibility is commonly approximated in the Lorentz model [8] by a density $\mathcal{N}$ of mechanical oscillators with mass m, charge q, spring constant k and viscous damping constant Γ. The dielectric function for a single-resonance is then [8, 9]

$$\varepsilon(\omega) = 1+4\pi\chi(\omega) = 1+\frac{\omega_p^{\ 2}}{\omega_0^{\ 2}-\omega^2+\mathrm{i}\gamma\omega}, \tag{1}$$

where $\omega_0 = (k/m)^{1/2}$, $\omega_p = (4\pi\mathcal{N}q^2/m)^{1/2}$ and $\gamma = \Gamma/m$. Eq. (1) is based on Newton's dynamical equation for the oscillator's displacement, x, caused by an applied electric field $E(t)$, namely

$$m\frac{\mathrm{d}^2x}{\mathrm{d}t^2}+\Gamma\frac{\mathrm{d}x}{\mathrm{d}t}+kx = qE(t)\ , \tag{2}$$

Many systems obey similar equations; the series LCR circuit of Figure 1 is of special interest here. In contrast to the mechanical oscillator, this circuit is treated using Kirchoff's potential law [10]. The result is an equation in charge, Q, *and* current, $\mathcal{J} = \mathrm{d}Q/\mathrm{d}t$, for the response to an EMF, $\mathcal{E}(t)$,

$$L\frac{\mathrm{d}\mathcal{J}(t)}{\mathrm{d}t}+R\mathcal{J}(t)+\frac{Q}{C} = \mathcal{E}(t)\,. \tag{3}$$

This is traditionally solved by introducing the complex impedance, $\hat{Z}$, and admittance, $\hat{Y}$. The result is

$$\hat{Y}_{\text{series}}(\omega) = \frac{1}{\hat{Z}(\omega)} = \frac{\mathcal{J}(\omega)_{\text{series}}}{\mathcal{E}(\omega)} = \mathrm{i}\omega\left[\frac{1/L}{\omega_{series}^2-\omega^2+\mathrm{i}\omega\gamma'}\right], \tag{4}$$

where $\omega_{\text{series}} = (1/LC)^{1/2}$ and $\gamma' = R/L$. The point to note is that the quantity in brackets, $\hat{Y}_{\text{series}}/\mathrm{i}\omega$, has the same form as $\chi(\omega)$ the material part of $\varepsilon(\omega)$. This similarity may be made explicit by substituting $\mathrm{d}Q/\mathrm{d}t$ for the current. Then the dynamical equation becomes the second-order differential equation

$$L\frac{\mathrm{d}^2Q}{\mathrm{d}t^2}+R\frac{\mathrm{d}Q}{\mathrm{d}t}+\frac{Q}{C} = \mathcal{E}(t)\,, \tag{5}$$

which is identical to Eq. (2), but with $Q(t)$ the analogue of $x(t)$. (See [11] for the quantum case.)

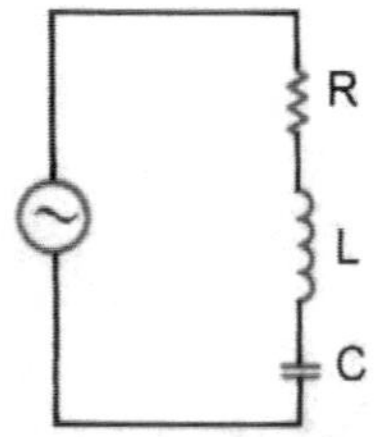

Figure 1. Analogue of $\chi(\omega)$.

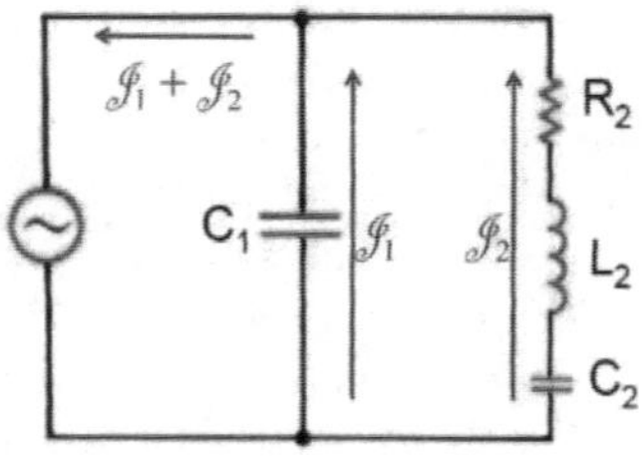

Figure 2. Analogue of $\varepsilon(\omega)$.

A full circuit analogue of $\varepsilon(\omega)$ is suggested by the rule that admittances of parallel circuits add. The result is the parallel combination of capacitive and LCR branches of Figure 2 with admittance

$$\hat{Y}_{\|}(\omega) = \hat{Y}_{\text{capacitor}}+\hat{Y}_{\text{series}} = \mathrm{i}\omega\left[C_1+\frac{1/L_2}{\omega_{\text{series}}^{\ 2}-\omega^2+\mathrm{i}\omega\gamma'}\right]. \tag{6}$$

C_1 has the value of the free-space susceptibility; the LCR branch that of matter susceptibility $\chi(\omega)$.

The ease of visualizing circuit currents suggests an extension of the analogy between $\hat{Y}_{\|}$ and $\varepsilon(\omega)$ to optical polarization currents. From Maxwell's equations [7] the displacement current density is

$$J_{\text{displacement}}(\omega) = \frac{1}{4\pi}\frac{\partial D}{\partial t} = \frac{1}{4\pi}\mathrm{i}\omega\varepsilon(\omega)E(\omega) = \mathrm{i}\omega\left[\frac{1}{4\pi}+\chi(\omega)\right]E(\omega)\,. \tag{7}$$

Here the current has been explicitly divided into its two parts:

1 A (virtual) vacuum-polarization current, $(1/4\pi)\ \partial E/\partial t$. The factor $\mathrm{i}\omega$ shows it leads the electric field by 90°. Its circuit analogue is the capacitive-branch current, $\mathcal{J}_1$.

2 A (real) material current associated with the rate of change of the polarization, $\partial P/\partial t$. This has the same frequency-dependent phase as $\mathrm{i}\omega\chi(\omega)$. Its circuit analogue is the series LCR branch current, $\mathcal{J}_2$.

Since the magnetic field of a propagating E-M wave arises from the displacement current, the frequency-dependence of optical properties is seen to result from interference between vacuum- and matter-polarization currents. This is a direct analogue of the frequency-dependent variations in external current for the circuit of Figure 2 caused by branch-current interference. Specifically:

a. At frequencies below resonance, $\omega << \omega_{series}$, both circuit branch currents are primarily "capacitive" and so in phase. [Capacitive impedance, $1/\omega C_2$, >> inductive impedance, ωL_2, in the LCR branch.] Similarly, in a dielectric-continuum, when $\omega \leq \omega_0$ both vacuum and matter displacement currents are largely in phase: Near resonance the matter leads; the *inertial* term in the equation for matter polarization is negligible compared to the spring's contribution, so the oscillator's motion is largely in phase with the non-inertial vacuum polarization over most of the range $\omega \leq \omega_0$.

b. At frequencies above resonance, $\omega \geq \omega_{series}$, inductance dominates the LCR branch. Its current lags, while that through C_1 still leads. Destructive interference determines the external current and, when the magnitude of the LCR branch current is large (small R_2) the external current may reverse. In the dielectric analogue, when $\omega \geq \omega_0$, *inertia* dominates the Lorentz-oscillator motion and the material current interferes *destructively* with the vacuum current. For large matter currents the *total* current may also reverse above resonance. This reverses the magnetic field of an E-M wave and propagation is inhibited. The result is semi-metallic reflectivity as exhibited by intrinsic silicon (reflectivity of 50 to 70% from 5 eV to 15 eV.) Figure 3 shows the reactive currents, phase and reflectivity for intrinsic silicon [12] and for a circuit with parameters matching silicon's, but with negligible resistance, $R_2 \sim 0$.

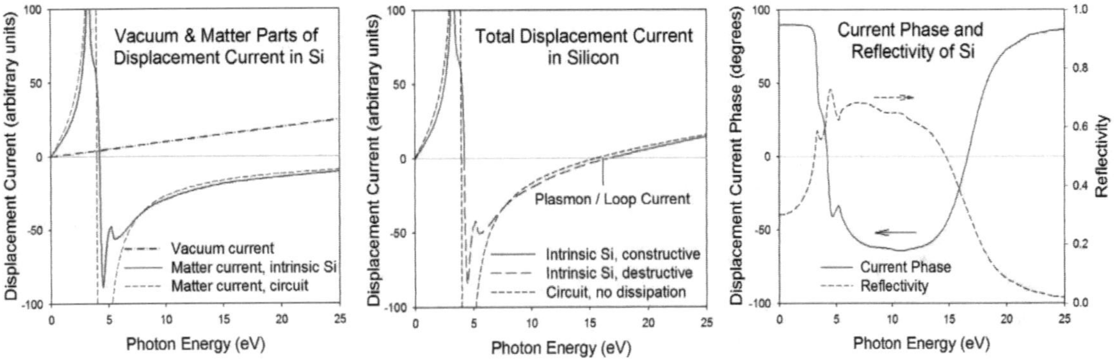

Figure 3. Displacement currents, phase and reflectivity for silicon, and currents for an analogous circuit model.

3. Excitations

The a-c circuit also provides insight into a dielectric's E-M excitations. Rearranging Eq. (6) yields

$$\hat{Y}_{\parallel} = \mathrm{i}\omega\left[C_1 + \frac{1/L}{\omega_{\mathrm{series}}^{\;2} - \omega^2 + \mathrm{i}\omega\gamma'}\right] \;=\; \mathrm{i}\omega C_1\left[\frac{\omega_{\mathrm{collective}}^{\;2} - \omega^2 + \mathrm{i}\omega(R_2/L_2)}{\omega_{\mathrm{series}}^{\;2} - \omega^2 + \mathrm{i}\omega(R_2/L_2)}\right] \qquad (8)$$

Here $\omega_{series} \approx (1/L_2C_2)^{1/2}$ and $\omega_{collective} \approx (1/L_2C_1 + 1/L_2C_2)^{1/2}$. Mathematically, the poles and zeros characterize $\hat{Y}_{\parallel}$ [13]. Physically they represent the extreme responses to electrical or optical probes.

a. Poles. The pole of $\hat{Y}_{\parallel}$ (an admittance maximum for $\omega > 0$) occurs at

$$\omega_{pole} = \sqrt{1/L_2C_2 - (R_2/2L_2)^2} + \mathrm{i}(R_2/2L_2)\,. \qquad (9)$$

This is a resonance in the LCR branch; it lies at $\omega_{series} \approx (1/L_2C_2)^{1/2}$ (for no dissipation). Both $\mathscr{I}_2$ and $\mathscr{I}_{external} = \mathscr{I}_1 + \mathscr{I}_2$ have maxima since $\mathscr{I}_1$ is negligible. Thus, a circuit pole represents the *single-branch* excitation of Figure 4. The dielectric-continuum analogue is a *single-particle* optical resonance at ω_0.

b. Zeros. The zero of $\hat{Y}_{\parallel}$ (impedance maximum for $\omega > 0$) occurs at

$$\omega_{zero} = \sqrt{(1/L_2)(1/C_1 + 1/C_2) - (R_2/2L_2)^2} + \mathrm{i}(R_2/2L_2)\,. \qquad (10)$$

Here, the *external current is zero* (minimum for real ω) regardless of EMF. This seemingly counter-intuitive condition involves a collective-resonance: a loop current circulating through *both* parallel branches, see Figure 5. This may be seen since ω_{zero} is the resonant frequency for the *series* combination of capacitive (vacuum) and LCR (matter) branches, $(1/L_2C_1+1/L_2C_2)^{1/2}$ (no dissipation).

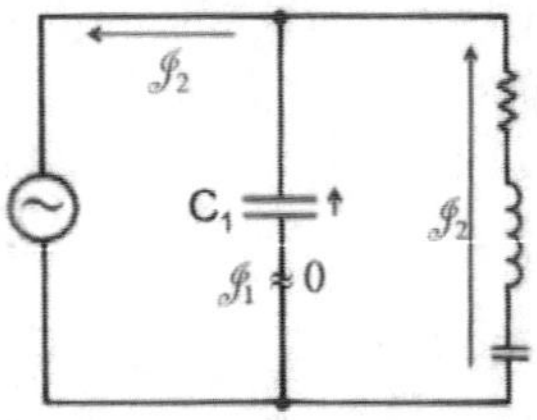

Figure 4. Single-branch resonance, the analogue of a single-particle resonance.

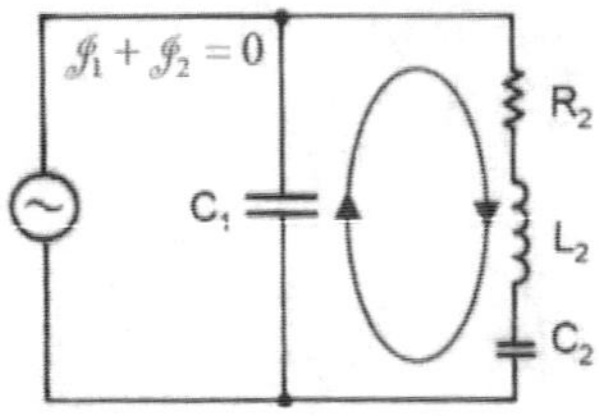

Figure 5. Circuit-loop resonance, the analogue of a plasmon resonance.

This resonance loop-current mode circulates *entirely* within the parallel network. It is not ususlly considered since an external voltage applied to the network terminals does not create an EMF within the loop that would excite the loop current (transients would dissipate). However, the mode can be easily excited by a changing magnetic field passing *through* the loop; say by moving a magnet into the loop. (Or, by coupling one of the elements to a signal. Cf. the tank circuit of a radio receiver [14, 15].)

Since $\hat{Y}_{||}/i\omega$ is the analogue of $\varepsilon(\omega)$, a zero of $\hat{Y}_{||}$ corresponds to a zero of the dielectric function, the hallmark of plasma modes [16]. Hence, the internal loop-current resonance is the circuit analogue of a bulk plasma excitation. The traditional picture of plasma excitations is of a fluctuation in an electron gas. The resulting charge imbalance yields a collective oscillation in the electron density. Such excitations are easily observed as quantized energy-loss by charged particles passing through condensed matter, especially in good metals and some semiconductors. However, in insulators the zeros of $\varepsilon(\omega)$ generally lie high in the complex-frequency plane. Then the plasmon lifetime is too short (line widths too broad) for the excitation to be well defined. The circuit analogue for this is a large value of R_2 that damps out any internal loop currents before a full oscillation takes place.

In summary, there are two distinct circuit excitations, the single-branch and loop resonances. Analogously, in a dielectric continuum there are also two distinct excitation classes: A Helmholtz decomposition [17] of the electric field yields a transverse component corresponding to optical excitations; and a longitudinal component corresponding to plasma modes.

4. Circuit sum rules analogies

We have use circuit analogies to clarify E-M processes in dielectrics. A reverse case is the prediction of integral circuit rules from optical sum rules: The *f* sum rule [4, 5] for electron density, $\mathcal{N}$, is

$$\int_0^\infty \omega\varepsilon_2(\omega)\,d\omega = 4\pi\mathcal{N}e^2\frac{\pi}{2m}\ . \tag{11}$$

This forms the basis of Smakula's equation [18] for defect density and for much quantitative optical spectroscopy. However, using the analogies between $\varepsilon(\omega)$ and $Y(\omega)$ and between m and L, one finds a similar rule for the dissipative part of the admittances of the circuits of the sorts considered above,

$$\int_0^\infty \mathrm{Re}\hat{Y}(\omega)\,d\omega = \frac{\pi}{2L}\ . \tag{12}$$

The physical basis of the *f* sum rule is the inertia of the optical response [19]. By analogy, the basis of the circuit rule is inductance. Remarkably, while both rules involve the dissipative response, they are *independent* of the dissipative parameters Γ and R, but depend only on m and L, respectively.

Acknowledgements

These studies were supported in part by the US Department of Energy, Office of Science, Office of Nuclear Physics under contract DE-AC02-06CH11357.

IOP Conf. Series: Materials Science and Engineering **80** (2015) 012012 doi:10.1088/1757-899X/80/1/012012

References

[1] Goldstein H 1950 *Classical Mechanics* (Addison-Wesley: Cambridge, MA)
[2] Wells D A 1938 *J. Appl. Phys.* **9** 312-320
[3] Louisell W 1964 *Radiation and Noise in Quantum Electronics* (New York: McGraw-Hill). Note: In this reference the inductance is often chosen to be unity. The resulting formulas do not contain *L*, which can be misleading.
[4] Bethe H A and Salpeter E E 1957 *Quantum Mechanics of One- and Two-Electron Atoms* (Berlin: Springer)
[5] Altarelli M, Dexter D L, Nussenzveig H M and Smith D Y 1972 *Phys. Rev.* B **6** [12] 4502-4509
[6] Bode H W 1945 *Network Analysis and Feedback Amplifier Design* (New York: Van Nostrand)
[7] Landau L D and Lifshitz E M 1960 *Electrodynamics of Continuous Media* (Oxford: Pergamon)
[8] Lorentz H A 1909 *The Theory of Electrons and Its Applications to the Phenomena of Light and Radiant Heat* (Leipzig: Teubner); reprinted 2011 (New York: Dover)
[9] Rosenfeld L 1951 *Theory of Electrons* (Amsterdam: North Holland)
[10] Beaty H W and Fink D G 2012 *Standard Handbook for Electrical Engineers* (New York: McGraw-Hill)
[11] Schiff L I 1955 *Quantum Mechanics* 2nd ed (New York: McGraw-Hill); see Sec. 35, pg. 253 for a discussion of the relation between current (momentum) and dipole matrix elements and their use in alternative treatments of optical transitions. In 3rd ed of 1968 see Sec. 44, pg 404.
[12] Shiles E (private communication); Smith D Y, Inokuti M, Karstens and Shiles E 2006 *Nucl. Instr. and Meth. in Phys. Res. B* **250** 1-5
[13] Churchill R V and Brown J W 1984 *Complex Variables and Applications* (New York: McGraw-Hill)
[14] Terman F E 1943 *Radio Engineer's Handbook* (New York: McGraw-Hill)
[15] Blanchard J 1941 *Bell System Tech. J.* **20** [4] 415-433
[16] Kittel C 1996 *Introduction to Solid State Physics* 7th ed (New York: Wiley)
[17] Arfken G B and Weber H J 1995 *Mathematical Methods for Physicists* 4th ed (San Diego: Academic Press) pp. 92-93
[18] Smakula A 1930 *Zeit. Physik* **59** [9 & 10] 603-614
[19] Altarelli M and Smith D Y 1974 *Phys. Rev. B* **9** [4] 1290-1298 and Altarelli M and Smith D Y 1975 *Phys. Rev. B* **12** [8] 3511

IOP Conf. Series: Materials Science and Engineering **80** (2015) 012013 doi:10.1088/1757-899X/80/1/012013

TL and EPR correlations in a quartz-like silicate mineral

Nilo F. Cano[1,*], Shigueo Watanabe[2]
[1]Departamento de Ciências do Mar, Universidade Federal de São Paulo, Av. Alm. Saldanha da Gama, 89, CEP 11030-400, Santos, SP, Brazil
[2]Instituto de Física, Universidade de São Paulo, Rua do Matão, Travessa R, 187, CEP 05508-090, São Paulo,SP, Brazil

E-mail: nilocano@if.usp.br

Abstract. A sample of silicate mineral considered as being *jadeite* but revealed to be more quartz-like, therefore named JQ2 has been studied concerning its TL and EPR properties. This study revealed a full correlation between the 210 °C TL peak and the g=2.0025 EPR signal. In both cases, the centers disappear at the same temperature and are restored after gamma irradiation. The TL peak at 210 °C may be related to the E'_1 centers formed on irradiation. JQ2 responds to very high γ-dose and can be used as a high-dose dosimeter using TL and/or EPR techniques.

1. Introduction
Synthetic as well as naturally available materials with luminescent properties have many applications such as radiation dosimetry and archaeological and geological dating [1-3]. The defect centers created by ionizing radiation are responsible for their TL properties [3]. The identification and characterization of these centers constitutes an essential step in understanding the mechanism of thermoluminescence (TL). In this context, electron paramagnetic resonance (EPR) provides a convenient and sensitive technique and it helps in providing support and further identification of the species detected by TL technique. The EPR technique has been extensively used in the identification of the main paramagnetic defects in the mineral silicates [2,3].
The dosimetric properties of these materials have already been studied using the TL and EPR techniques and it was suggested that they show potential for use in high-dose dosimetry [4,5]. Rocha et al. [6] studied the natural jade using the thermally stimulated exoelectron emission technique. Arslanlar et al. [7] investigated some physical properties dependent on point defects of natural crystals of jade from Turkey using the cathodoluminescence, radioluminescence and TL techniques; but there is no evidence in the literature describing the defects responsible for the TL properties in quartz-like and the correlation between TL and EPR measurements. In the present work, green samples acquired as *jadeite*, have been investigated concerning the EPR and TL properties and also the effects of the ionizing radiation and the thermal treatments at high temperatures of the samples, elucidating the defect center responsible for production of 210 °C TL peak in natural quartz-like.

2. Materials and experimental

The sample studied here was acquired from LEGEP Minerals Ltd. in São Paulo. It presents a green color. About 30 grams of the sample were pulverized and sieved to retain grains sizes between 0.080 and 0.180 mm for TL and EPR measurements. Those with size smaller than 0.080 mm have been used for X-ray fluorescence (XRF) measurement.

The TL measurements have been carried out on Harshaw model 4500 TL reader (temperature range: 50 °C up to 400 °C and heating rate: 4 °C/s). The EPR measurements were performed using a Bruker EMX spectrometer using a rectangular cavity (ST ER4102) with a microwave frequency of 9.767 GHz.

For irradiation, IPEN/CNEN/SP - Institute for Energy and Nuclear Researches Radiation Center's ^{60}Co source was used. The XRF measurement was carried out at Department of Mine and Petroleum, University of Sao Paulo.

3. Results and discussion

An analysis of the main components of the quartz-like (JQ2) sample was obtained by XRF. Results are presented in Table 1. This analysis was performed to identify the chemical elements in the sample, and for future studies about which of these elements are responsible for the TL signal. Although the XRF results have shown that has the same compound that the quartz crystal, the TL glow curve of α-quartz is different, when compared with the TL glow curves of JQ2 sample which is described by Barbosa et al.[8].

Table 1 Weight % of basic oxides components and impurities in JQ2 determined using X-ray fluorescence analysis.

Sample	Compound (w. %)									
	SiO_2	Al_2O_3	Fe_2O_3	Cr_2O_3	MgO	CaO	Na_2O	K_2O	TiO_2	P_2O_5
JQ2	87.2	4.55	0.14	0.40	0.37	0.16	0.19	1.48	0.01	0.04

The TL glow curves of this sample as function of high dose gamma irradiation have been reported by Barbosa et al. [8]. Here, we studied the correlation between the TL glow curves of each peak and the EPR signals.

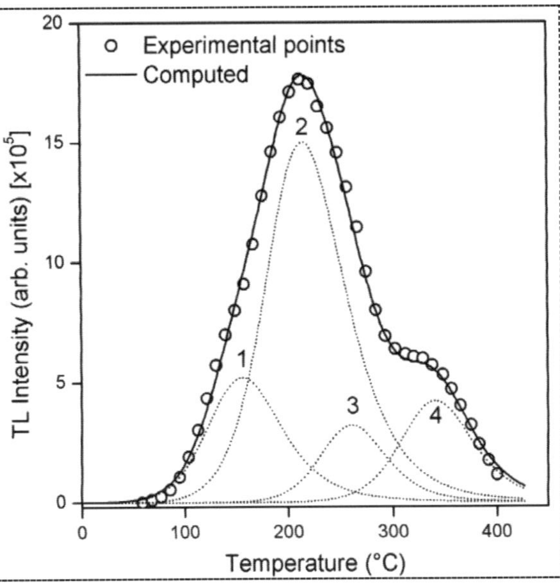

Figure 1. TL glow curve from natural JQ2 pre-annealed at 500 °C/30min. and subsequently receiving a γ-ray dose of 10 kGy. A good fit between the experimental glow curve (circles) and the simulated glow curve (full line) can be acheived by assuming the presence of four peaks.

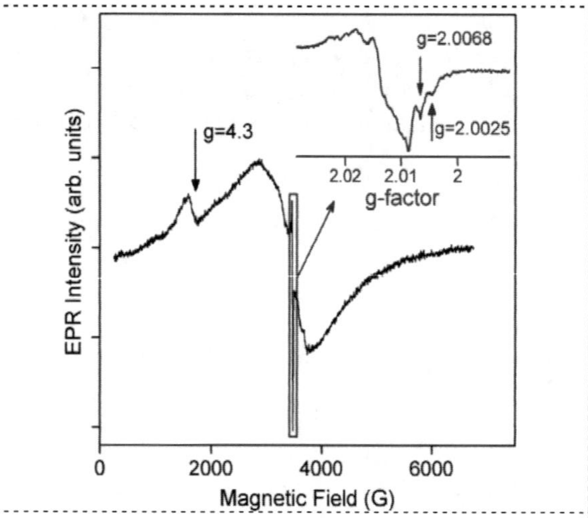

Figure 2. EPR spectrum of non-irradiated JQ2. In the inset of the figure, detail of the EPR spectra in the region of the E´$_1$ center.

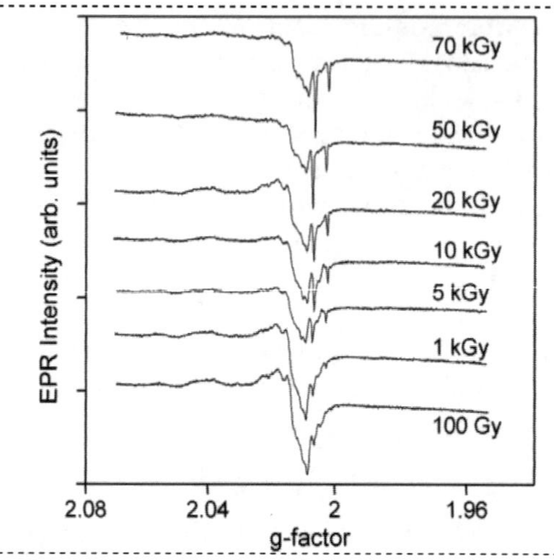

Figure 3. EPR spectrum of JQ2 irradiated with different gamma doses.

In this work, EPR studies were carried out mainly to identify the defect centers responsible for the observed TL peaks in JQ2. Fig. 2 shows the EPR spectrum for JQ2 sample without irradiation. It shows g=4.5 signal due Fe^{3+} ion and an intense signal extending from about 2200 to 5000 G due to spin-spin interaction of Fe^{3+} ions. In the inset, the EPR signal with g between 2.00 and 2.02 are shown. We then irradiated JQ2 sample with gamma rays with dose varying from 100 Gy and 70 kGy. The EPR results are shown in Fig. 3 for g-values from 1.96 and 2.08.
Analysing Fig. 3, we can see two EPR signals at g = 2.0025 and at g = 2.0068. They are found to increase with irradiation dose. To see whether there is any correlation between EPR signals and the TL peaks behaviour, we compared EPR results with TL intensity behavior of TL peaks obtained by Barbosa et al. [8]. Fig. 4 shows such comparison.

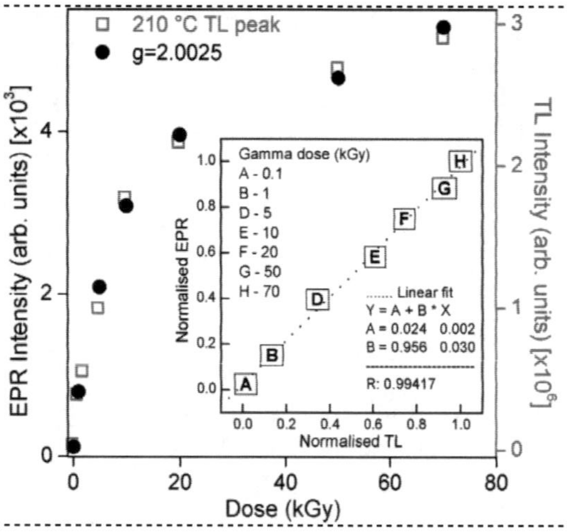

Figure 4. Correlation between the EPR signal at g=2.0025 and the 210 °C TL peak behavior under gamma dose irradiation. In inset of the figure, the linear dependence between the E´1 center and 210 °C TL peak.

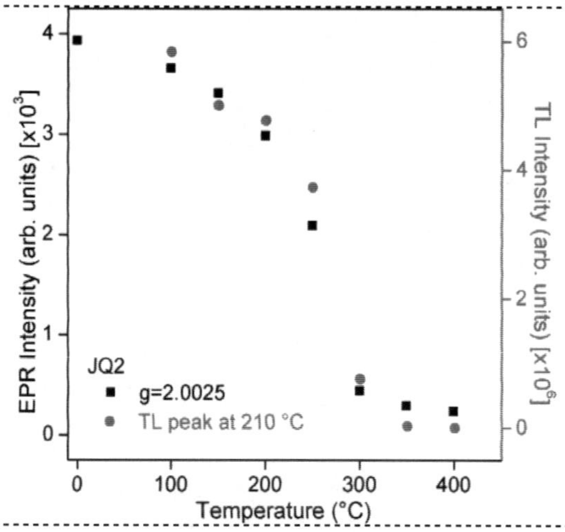

Figure 5. Decay of the 210 °C TL peak and the g=2.0025 EPR signal of the JQ2 samples submitted to different thermal treatments.

Fig. 4 shows a clear correlation between the EPR and TL results, i.e., the EPR signals increase parallel to 210 °C TL peak intensity with irradiation dose. In the inset of Fig. 4 presents the linear dependence between TL and EPR results indicates that the growths of both the 210 °C TL peak and EPR center with gamma-dose are similar. It leads us to say that the center responsible for the EPR signal at g=2.0025 is the same as the one that produce 210 °C peak, since both signals increase in the same way with dose. Therefore, it may be concluded that the center responsible for the EPR signal at g=2.0025 is the same as the center that produces the 210 °C TL peak. The EPR signal observed at g=2.0025 has been assigned to E'_1 type centers. This center grows with radiation dose. Some other silicates, as petalite [9], kaolinite [10] and diopside [11], have shown the formation of E'_1 centers on irradiation.
The isochronal thermal decrease of EPR signals of the JQ2 sample were performed after sample irradiation with gamma-ray dose of 5 kGy and annealed from 100 up to 400 °C. The EPR signal at g~2.017 do not change, while g=2.0068 and g=2.0025 signals decrease with increasing annealing temperature. The 210 °C TL peak behavior was examined under similar annealing and irradiation conditions. Fig. 5 shows a comparison of these results and it is observed that the EPR signal due to E_1'-center and TL peak at 210 °C behave in a very similar manner. The observed similarity among TL and EPR results indicates their clear correlation. The centers disappear at the same temperature and the gamma irradiation restores the centers simultaneously in the same way. Hence this glow peak may be attributed to the recombination of electrons that are liberated from electron traps with the hole centers.

4. Conclusions

The EPR results of JQ2 presented the typical g≅4.5 Fe^{3+} signal around 1500 G, and besides, an EPR signal around g≅2.0 which also due to Fe^{3+} ion. The signals around g=2.0025 and g=2.0068 are assigned to E_1'-center and peroxy center, respectively. These signals grow with radiation dose and the signal due to E_1'-center decrease with temperature. Data correlation analysis of EPR and TL results showed that the EPR signal at g=2.0025 and the TL peak at 210 °C can be attributed to the same defect. In both cases, the centers disappear at the same temperature and are restored after gamma irradiation. Even though the presence of Fe impurity ions was identified in the matrix of JQ2 crystals through XRF and EPR studies, they do not take part in the TL emission process directly. They induce the formation of native lattice defects that act as donors and acceptors. One needs further data to present a complete model about these defect centers for the TL glow curve of JQ2.

Acknowledgments
This work was carried out with financial support from FAPESP.

References

[1] Ayta WEF, Silva VA, Cano NF, Silva MAP and Dantas NO 2011 *J. Lumin.* **131** 1002
[2] Cano NF, Yauri JM, Watanabe S, Mittani JCR and Blak AR 2008 *J. Lumin.* **128** 1185
[3] Cano NF, Turbiani-Filho IT, Gennari RF, Munita CS, Souza MC, Ângulo RJ and Watanabe S 2013 *Quatern. Int.* **306** 137
[4] Melo AP, Valerio MEG and Caldas LVE 2004 *Nucl. Instrum. Meth. B* **218** 198
[5] Teixeira MI, Melo AP, Ferraz GM and Caldas LVE 2010 *Appl. Radiat. Isotopes* **68** 582
[6] Rocha FDG, Cecatti SGP and Caldas LVE 2002 *Radiat. Prot. Dosim.* **100** 417
[7] Arslanlar YT, Garcia-Guinea J, Kibar R, Çetin A, Ayvacıklı M and Can N 2011 *Appl. Radiat. Isotopes* **69** 1299
[8] Barbosa RF, Reichmann F, Cano NF, Rocca RR and Watanabe S 2013 *Phys. Status Solidi* C **10** 242
[9] Souza SO, Chubaci JFD, Selvin PC, Sastry MD and Watanabe S 2002 J. Phys. D Appl. Phys. **35** 1562
[10] Ildefonse P, Muller JP, Clozel B, Calas G 1991 *Mater. Res. Soc.* **212** 749
[14] Cano NF, Yauri JM, Watanabe S, Mittani JCR and Blak AR 2008 *J. Lumin.* **128** 1185

Optical Absorption and Electron Paramagnetic Resonance studies of two different solid solutions of Pyralspite Garnet

S Watanabe[1], **S R Q Espinoza**[1], **J F D Chubaci**[1], **N F Cano**[2], **D R Cornejo**[1]
[1]Institute of Physics, University of São Paulo, São Paulo, Brazil
[2]Federal University in State of São Paulo, Campus-Santos, Brazil

E-mail: nilocano@if.usp.br

Abstract. Two different solid solutions of pyralspite garnet have been investigated as to their optical absorption and EPR properties. The absorption band around 9850 cm^{-1} is due to Fe^{2+}. After heating above 950 °C we found this band diminishes considerably; which was interpreted as Fe^{2+} loosing an electron to become Fe^{3+}. The EPR spectrum of sample consisted of a straight line with angular coefficient of about 176°. After 600°C/1hr annealing, the spectrum starts deviating from a straight line. A strong typical EPR signal is observed around g = 2.0 after annealing at 850°C. We assume that with high temperature annealing a large number of Fe^{2+} are converted to Fe^{3+} as the optical absorption has shown. These results were also confirmed by chemical reaction. The susceptibility vs. magnetic field measurement has shown that the samples annealed at temperatures below 850°C present normal paramagnetic behavior, however, annealed above 900 °C, they show hysteresis, namely ferromagnetic behavior.

1. Introduction

The Garnet Group is composed of six silicate minerals. They are pyrope, almandine, spessartine, grossular, andradite and uvirovite. The three first ones belong to the subgroup called *pyralspite* and have the general formula $X_3Al_2Si_3O_{12}$ and the three others to the subgroup called *ugrandite* and have the general formula $Ca_3Y_2Si_3O_{12}$. In the nature it is not common to find pure pyrope or almandine or spessartine rather solid solutions of them. Since grossular has formula $Ca_3Al_2Si_3O_{12}$, very often it is found as one of the compounds in the solid solutions. In the present work we investigated two solid solutions of piralspite subgroup elements as to their optical absorption (OA) and EPR behavior.
Two mains works on OA spectroscopy of garnets are due to Manning [1] and Slack & Chrenko [2]. They list 13 absorption bands between 4300 cm^{-1} and 27300 cm^{-1}, a large part of these bands is due to Fe^{3+} in a octahedral environment. A huge band around 7700 cm^{-1} and smaller 5900 and 4200 cm^{-1} were assigned to Fe^{2+}. The bands at 23500 and 24500 cm^{-1} were attributed to Mn^{2+}.

2. Materials and methods

One solid solution containing 79.95% almandine, 14,28% pyrope, 3.52% spessartine and 2.34% grossular and the other one containing 54.45% almandine, 43.6% spessartine, 1.2% pyrope and 5.5% grossular were acquired for the present work. 5mmx 5mm x1.5 mm slabs have been prepared for optical absorption (OA) measurements using CARY spectrophotometer. For EPR measurements the samples were crushed and sieved retaining grains with 0.080 to 0.180 mm size. A Bruker EPR spectrometer in X-band was used for the measurements. As part of an important experiment the samples were annealed at high temperatures of 800 to 1100°C before irradiation.

3. Results

Fig. 1 shows the OA spectrum of the sample 2 containing large almandine and spessartine components. Fig. 2 presents the OA spectrum of the sample 1 with a large concentration of almandine, subjected to annealing at 300, 500, 700 and 900°C for one hour. The optical absorption spectrum came out to be quite complex with several absorption bands in the range from near ultraviolet to near infrared. All these bands have been identified by comparison with absorption spectrum of almandine measured and identified by Manning [1] and also with the absorption spectrum of spessartite by Slack and Chrenko [2]. The irradiation with gamma-rays even with very high dose did not change the absorption spectrum. In first place, the background absorption increases such that starting around 900°C heating the bands in visible range disappear; secondly the Fe^{2+} band around 1200 nm decreases indicating that Fe^{2+} liberates electrons and Fe^{3+} ions are the result.

Fig.3 shows EPR spectrum of natural spessartite-almandine sample, which consist of quasi-linear with a slope of defined by un angle θ=102.5° with respect to horizontal H-axis. It extend from H=0 to H=6000 Gauss. Full line is the experimental one and dotted one calculated result, using:

$$f(x) = \frac{A}{\sigma\sqrt{2\pi}} e^{-\frac{(x-x_0)}{2\sigma^2}} \quad (1)$$

$$\frac{df(x)}{dx} = \frac{A(x-x_0)}{\sigma^3\sqrt{2\pi}} e^{-\frac{(x-x_0)^2}{2\sigma^2}} \quad (2)$$

with (1) as the absorption line and (2) the first derivate EPR line .

We then carried out a series of measurements keeping 10 mW microwave power and submitting the samples to heat treatment from 400 to 1000°C. At 400°C annealing the EPR straight line signal present a plateau from 3500 to 4200 Gauss; at 700°C annealing gives a sharp decrease of signal around 3500 Gauss occurs with the intensity varying from 45 ~ 45 to -125 and then a plateau follows extending to 4800 Gauss; for 800 to 850°C annealing the signal is constant between 1000 at 3200, 3400 Gauss, at this point the EPR intensity drops; to -500 and then gradually increases becoming constant from around 4800 Gauss on above 900 to 950°C the signal now has normal form around g=2.0. However the EPR intensity increases such that it varies from +1200 to -1500. Annealing the sample to higher temperature produces a signal with intensity varying from +3500 to -3000 a really gigantic signal. This is related to the fact at this temperature Fe^{2+} liberates electrons and becomes Fe^{3+}, which contributes through magnetic dipole-magnetic dipole interaction to this large EPR signal.

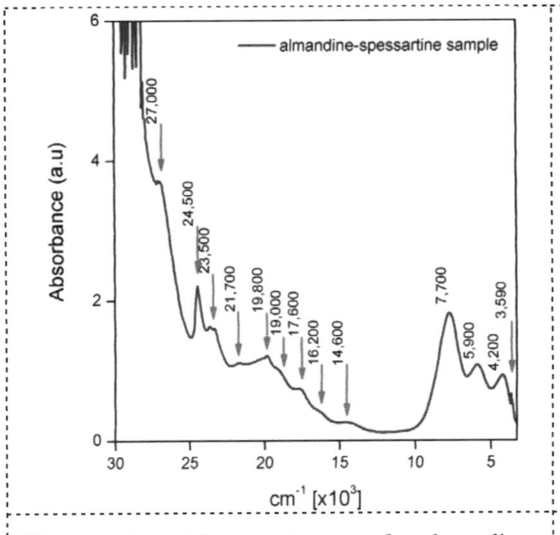

Figure 1. AO spectrum of almandine-spessartine sample.

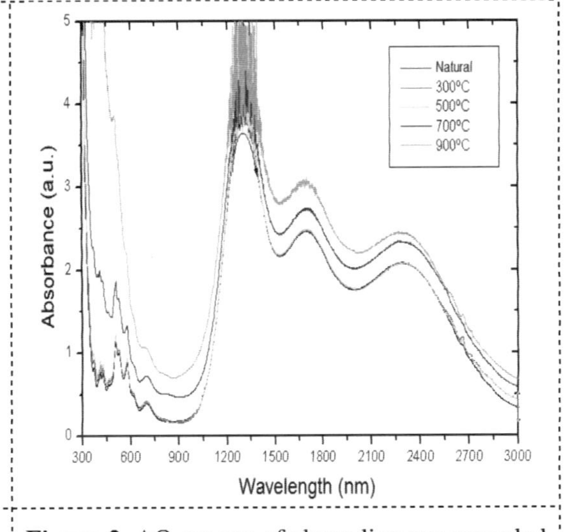

Figure 2. AO spectra of almandine pre-annealed a 300, 500, 700 and 900 °C.

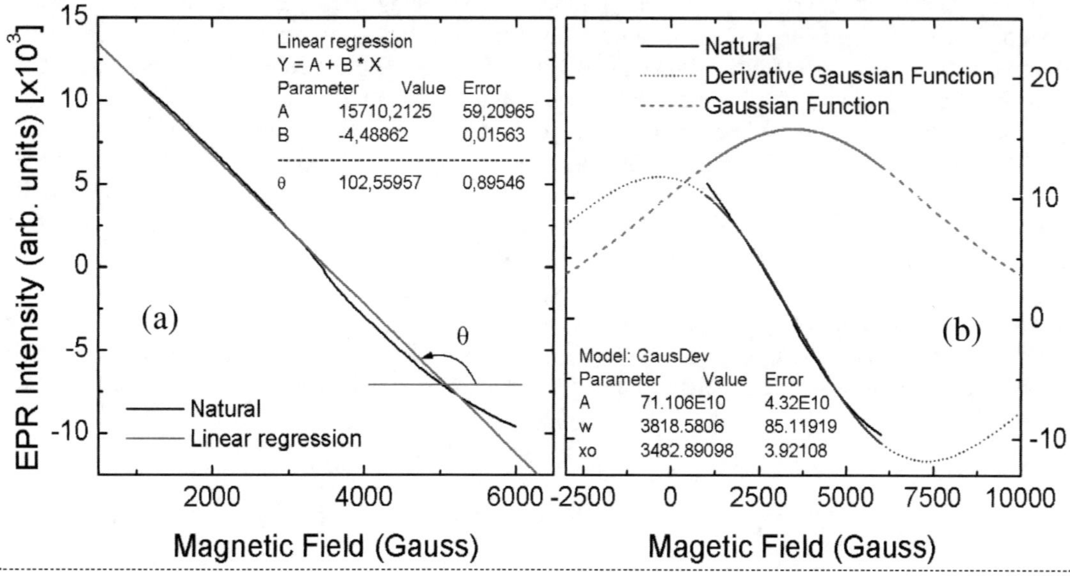

Figure 3. EPR spectra of Spessartite-Almandine natural non-heated sample. (a) sample natural and (b) calculated Gauss function and derivate Gauss function. The microwave absorption curve (----) and its first derivate (----) are calculated from eq. (1) and eq. (2).

Fig. 4 shows how the spectrum of Fig. 3 changes with annealing samples from 600 to 1100°C annealing.

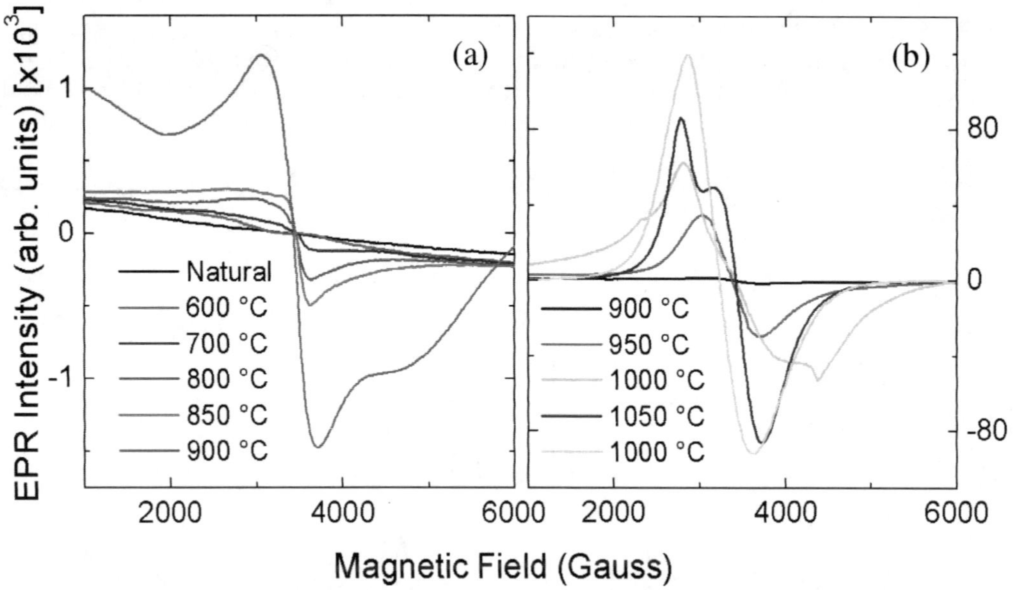

Figure 4. The EPR intensity for different annealing temperature. (a) EPR spectra of 600 to 900°C; that of untreated natural sample is shown for comparison. In (b), EPR spectra of 900 to 1100°C

Fig. 5 shows the paramagnetic behavior of the natural sample and that annealed at 800, 900, 1000 and 1100°C for one hour. The 900°C annealed sample presents a beginning of hysteresis behavior at low field region. The sample annealed at 1000 and 1100°C show stronger hysteresis behavior. Coercitivity of 250 to 290 Oe was observed. This result is compatible with ferro or ferrimagnetism.

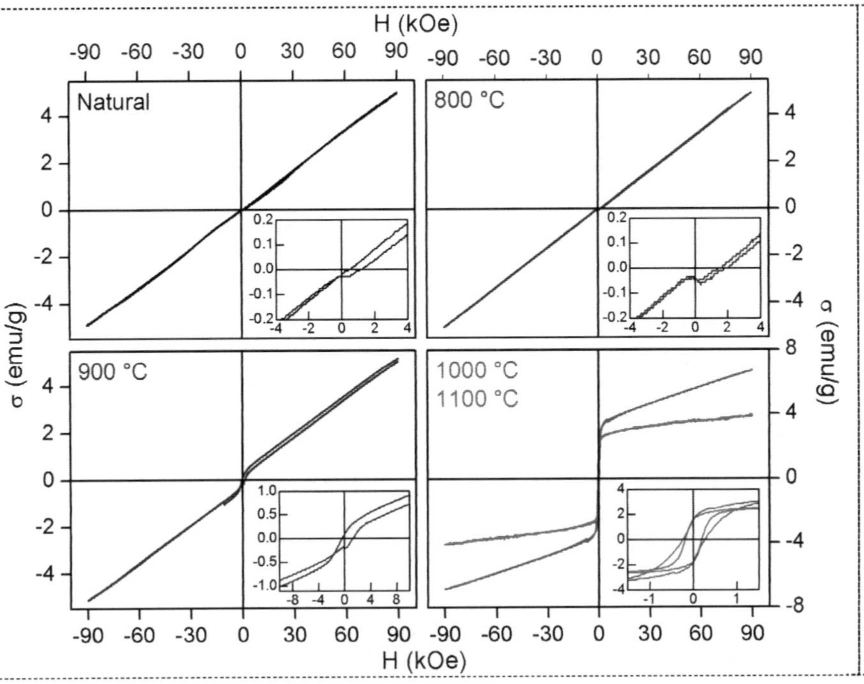

Figure 5. the paramagnetic behavior of the natural sample and that annealed at 800, 900, 1000 and 1100°C for one hour.

4. Conclusions

Several absorption bands in the near UV, visible and near IR at 330, 410, 450, 500, 1300, 1700 and 2600 nm, are observed. The color of natural sample is due to the 410, 450 and 500 nm bands. 410, 500, 1300 and 1700 nm bands have been attributed, respectively, to Fe^{3+}, Mn^{2+}, Fe^{2+} and Fe^{2+}.
The irradiation and annealing did not affect these absorption bands, however, heating above 850 - 900°C temperature, the background absorption increase such that the crystal becomes darker.
The EPR spectrum of untreated natural spessartite-almandine is practically a straight line with 102° inclination with respect to horizontal (H-field) axis.
With the annealing above 800 °C the Fe^{2+} band starts diminishing, meaning Fe^{2+} emits electron and becomes Fe^{3+}. On the other band, as shown in Fig. 4 the signal around g=2.0 due to Fe^{3+} increases with heat treatment above 900°C. We assume that with high temperature annealing large number of Fe^{2+} is converted to Fe^{3+} as the optical absorption has shown. These results were also confirmed by chemical reaction. The susceptibility vs. magnetic field measurement has shown that the samples annealed at temperatures below 850 °C present normal paramagnetic behavior, however, annealed above 900°C, they show hysteresis, namely ferromagnetic behavior.

Acknowledgments
This work was carried out with financial support from FAPESP.

References

[1] Manning P 1967 *Canadian Mineralogist* **9** 237
[2] Slack GA, Chrenko RM 1971 *J. Optical. Soc. Am.* **61** 1325

Temperature dependent luminescence characteristics of $KBe_2BO_3F_2$ and $RbBe_2BO_3F_2$

J Martincik[1,2], V Babin[1], L Liu[3], X Wang[3], C T Chen[3], A Beitlerova[1], E Mihokova[1], and M Nikl[1]

[1] Institute of Physics AS CR, Cukrovarnicka 10, Prague, Czech Republic
[2] FNSPE, Czech Technical University, Prague, Czech Republic
[3] Beijing center for crystal research and development, Technical Institute of Physics and Chemistry, Chinese Academy of Sciences, P.O. Box 2711, Beijing, China

E-mail: jiri.martincik@fjfi.cvut.cz

Abstract. This paper reports on a study of the luminescence characteristics of $KBe_2BO_3F_2$ (KBBF) and $RbBe_2BO_3F_2$(RBBF) crystals in UV/visible spectral range. The KBBF crystals are very popular for their nonlinear optical properties, however they have a potential to be used as scintillators for neutron detection. To determine the effectiveness of KBBF scintillation we combine the results from measurements of optical absorption; radioluminescence; light yield; photoluminescence and decay kinetics in the temperature range 8-330 K. Temperature dependence of luminescence in KBBF crystals is discussed.

1. Introduction

The $KBe_2BO_3F_2$ (KBBF) crystals have been known since 1968 and today they are very popular for their nonlinear optical properties and large band gap enabling their application for frequency multiplication down to VUV spectral region, see e.g. [1, 2]. The large band gap of the host KBBF can incorporate many rare earth ions to alter its luminescence properties for applications such as lasers or scintillators [3]. Research of borate compound system is very intensive and many new structure types with rare earth elements were discovered [4].

The investigation of various borate crystals [5, 6, 7] shows that these materials are suitable for neutron detection with efficiency comparable with other systems used for neutron detection [8].

Our research is focused on examining the effectiveness of scintillation of KBBF crystals in order to find a suitable candidate for scintillator for neutron detection. Our interest is the luminescence of defects and its temperature dependence in the UV/visible part of spectra. Only few papers dealing with investigation of the luminescence characteristics of KBBF are published [9].

2. Experimental details

KBBF and $RbBe_2BO_3F_2$ (RBBF) crystals were grown by a localized spontaneous nucleation technique without seeding by the flux method [10]. Polycrystalline KBBF powders were mixed with KF and B_2O_3 in a proper ratio and melted at around 750 °C. The crystals were grown at a cooling rate of 1.0-3.0 °C/d. The cooling range was more than 60 °C and the growth period was about two months. The growth procedure was similar for RBBF crystal growth by using a NaF-B_2O_3 flux system. These crystals could also be grown by the hydrothermal method. The growth details can be found in [11, 12]. After the growth had been finished, the crystals were cut into small pieces and optically polished for luminescence measurements. Samples are cut into sizes of about 5×5 mm^2 with thicknesses varied from 0.2 mm to 1.5 mm. At room temperature, the KBBF crystals are chemically stable and nonhygroscopic with the density of about 2.4 g·cm^{-3}. Eleven samples grown by different methods are prepared and optically polished as listed in table 1.

Table 1. Composition of KBBF and RBBF crystals; h-means hydrothermal-grown; f-means flux-grown.

Number	1	2	3	4	5	6	7-11
Composition	h-RBBF	h-RBBF	h-KBBF	f-KBBF	f-KBBF	f-RBBF	f-KBBF
Growth temp. (°C)	430	430	450	720	720	740	700
Flux system	Rb_2CO_3	Rb_2CO_3	KOH	KF-B_2O_3	KF-B_2O_3	NaF-B_2O_3	KF-K_2O-B_2O_3

Optical absorption spectra of all samples were measured at spectrometer Shimadzu 3101PC. Radioluminescence, photoluminescence and decay kinetics were measured by the custom made spectrofluorometer Horiba Jobin Yvon 5000M with a photon counting detector TBX-04 (IBH Scotland). The steady-state emission and excitation spectra of the KBBF were measured at the same conditions in the temperature range 8-330 K. The measurements were performed in a closed cycle cryostat (Janis). All spectra were corrected for instrumental distortions.

Luminescence decay kinetics in the microsecond time range was measured at 8-330 K by the same set-up under excitation by a xenon flash lamp FX-1152 (EG&G) and using the multichannel scaling method. A deconvolution procedure was used to extract the true decay time data using the SpectraSolve software package (Ames Photonics).

Light yield under an alpha particle excitation from ^{241}Am radioisotope was measured by the method of pulse-height spectra. Using a silicon grease, the sample is optically coupled to a hybrid photomultiplier (HPMT) [13] model DEP PPO 475B. Signal from HPMT is processed by spectroscopy amplifier ORTEC 672 and multichannel buffer ORTEC 927TM. Pulse-height spectrum is displayed on PC. All measurements were performed at room temperature.

3. Results and discussion

Our results combine the measurements of absorption, radioluminescence and photoluminescence emission and excitation spectra. We mainly focus on the luminescence band around 350 nm and its temperature dependence within 8-330 K. Furthermore, the decay kinetics of emission transitions in visible range in KBBF crystals was also studied.

Radioluminescence spectra in figure 1 are very similar to those of photoluminescence presented in [9], peaking at about 310-320 nm (4 eV). In some samples (RBBF-1, RBBF-6) their integral efficiency is quite high, about 300-400 % of that of $BiGe_3O_{12}$ (BGO) standard scintillator sample. However, light yield measurements of two most intensely emitting samples, no. 1 and 6 provided the values which are approximately an order of magnitude lower than those of the BGO standard scintillator. This result points to intense slow decay components in the KBBF scintillation response. This assessment is supported by strongly increasing LY value with increasing shaping time and by the presence of slow decay components in photoluminescence decays down to tens of microseconds. The results are reported in table 2.

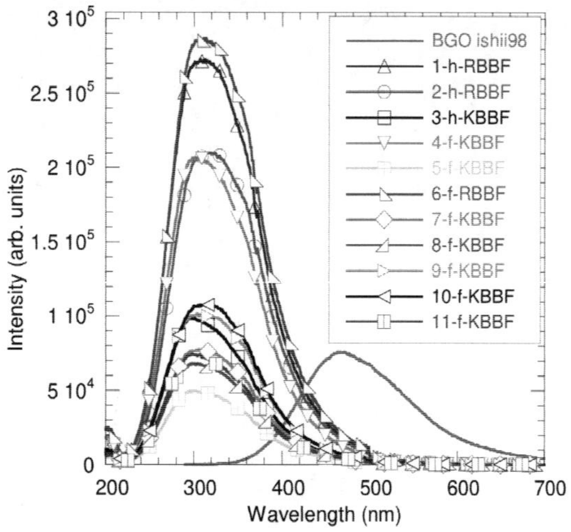

Figure 1. Radioluminescence spectra (excited by X-ray source, 40 kv, 15 mA).

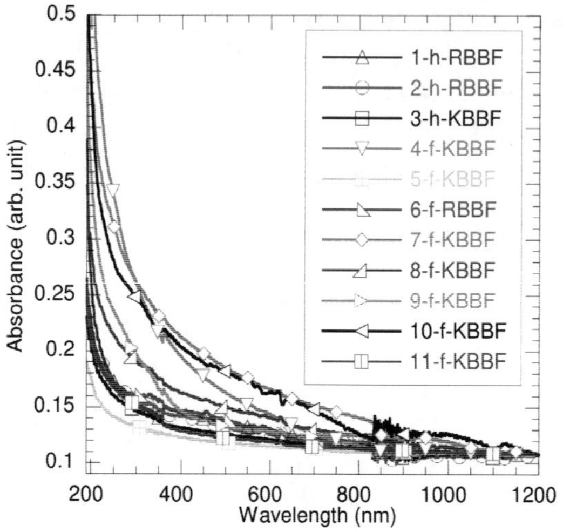

Figure 2. Optical absorption spectra of the sample set.

Slowly increasing absorbance in figure 2 towards shorter wavelengths can be due to scattering surface or any scattering inclusions in the samples without any apparent absorption bands. From the point of view of light scattering KBBF no. 5 is the best (least scattering).

Table 2. Scintillation properties of h-RBBF-1 and f-RBBF-6 crystals. Pulse height spectra were excited by α-particles from ^{241}Am source with thin Palladium layer at various shaping time from 0.5 to 10 μs. Number of photoelectrons (N_{phels}/MeV) was calculated from used energy of ^{241}Am with layer 4.8 MeV)

	h-RBBF-1					f-RBBF-6				
Shaping time (μs)	0.5	1	2	3	6	0.5	1	2	3	6
N_{phels}	30	28.6	34.5	39.3	56	32.3	42.3	51.6	51	90.5
FWHM (%)	34.3	43.7	≈30	45.9	39.8	36.8	≈40	≈30	≈30	≈30
LY per MeV (N_{phels}/1MeV)	6.25	5.96	7.19	8.19	11.67	6.73	8.82	10.75	10.63	18.85

To deeper investigate the luminescence of defects we measured the decay kinetics and temperature dependence of the luminescence band peaking at about 350 nm.

The main interest was focused on both the changes of emission intensity and shortening of decay components with increasing temperature. Temperature dependence of emission characteristics was measured in the range 8-330 K.

Temperature dependence of photoluminescence spectra is shown in figure 3. There seem to be a slight shift of the position of the emission maximum between the low and intermediate temperatures pointing to the composed character of the PL spectrum. This observation was further pursued by measurement of PL decays.

Photoluminescence decays were not single exponential and could only be fit by the sum of several exponentials. Figure 4 presents an example of the photoluminescence decay fit by three exponential function. The slowest component is very weak, therefore two exponential functions were considered for the fits of temperature dependence of PL decays.

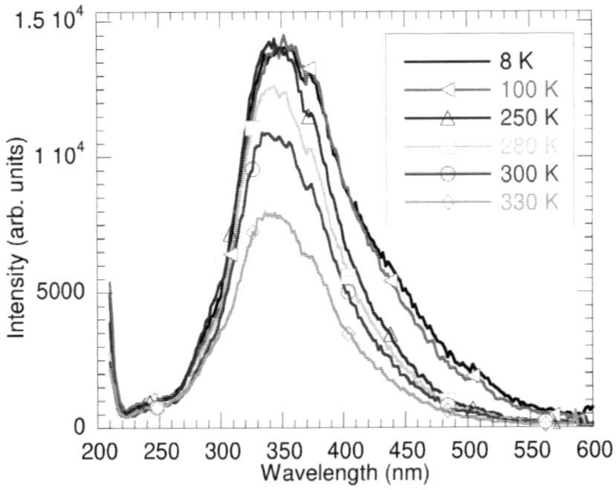

Figure 3. Photoluminescence spectra for exc=195 nm of sample no. 2.

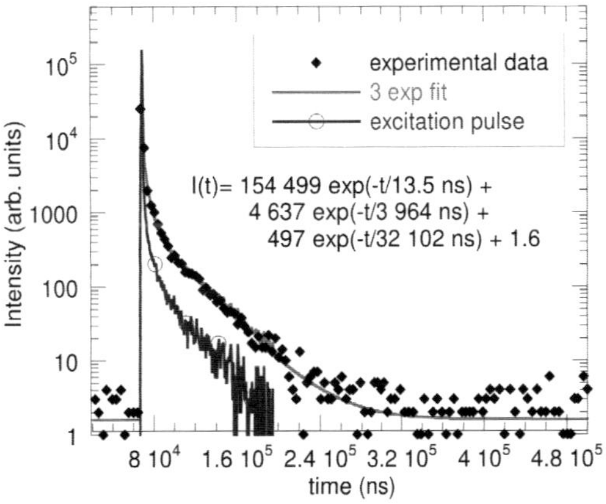

Figure 4. Photoluminescence decay under exc=195 nm and em=237 nm of sample no. 2. Solid line is the approximation made by convolution of instrumental response to excitation pulse and function I(t).

The decay times obtained from two-exponential approximation of PL decays as a function of temperature are shown in figure 5.

The two components of the decay indeed do not seem to be associated with the same luminescence center. This feature confirms a composed character of the PL spectrum. The behavior of fast decay can be satisfactorily described by a phenomenological model of the luminescence center consisted of three excited state levels (see the scheme in figure 6a). Observed decay corresponds to the slowest decay component of the luminescence center. Similar model was used in [14].

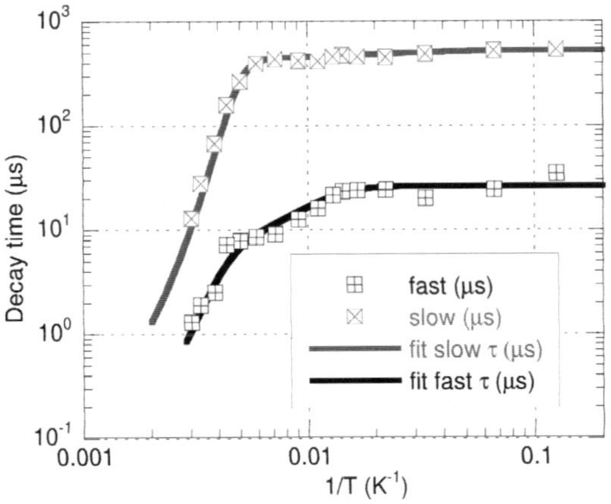

Figure 5. Temperature dependence of decay times of sample no. 2, exc=195 nm, em=350 nm; temperature range 3-330 K. Squares are experimental data, solid lines are the fits using models desribed in the text.

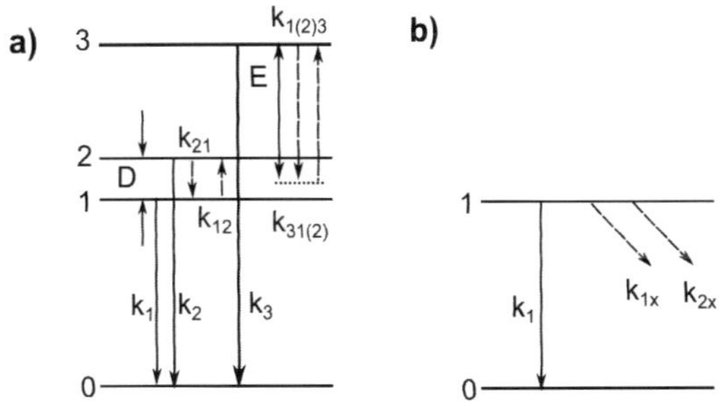

Figure 6. Energy level diagrams used for the description of decay kinetics. a) fast decay component model; b) slow decay component model

IOP Conf. Series: Materials Science and Engineering **80** (2015) 012015 doi:10.1088/1757-899X/80/1/012015

According to such model the shortening of the decay time with increasing temperature is caused by thermally stimulated non-radiative transitions between the excited state levels as depicted in the scheme in figure 6a. The time evolution of the populations N_1, N_2 and N_3 of the excited levels 1, 2 and 3, respectively, can be described by the following rate equations:

$$\begin{aligned}\frac{dN_1}{dt} &= -k_1N_1 - k_{12}N_1 - k_{13}N_1 + k_{21}N_2 + k_{31}N_3\\ \frac{dN_2}{dt} &= -k_2N_2 - k_{21}N_2 - k_{23}N_2 + k_{12}N_1 + k_{32}N_3\\ \frac{dN_3}{dt} &= -k_3N_3 - k_{31}N_3 - k_{32}N_3 + k_{13}N_1 + k_{23}N_2\end{aligned} \tag{1}$$

$$k_{21} = K\,(n+1)\,,\ k_{12} = Kn,\ n = \frac{1}{exp(\frac{D}{k_bT}) - 1} \tag{2}$$

$$k_{31} = K^{'}\left(n^{'}+1\right),\ k_{31} = K^{'}n^{'},\ n^{'} = \frac{1}{exp\left(\frac{E}{k_bT}\right) - 1} \tag{3}$$

Here K, n and D are the zero-temperature transition rate between the levels 1 and 2, the Bose-Einstein factor and energy spacing between the levels, respectively.

The parameters of the fit shown as a black solid line in figure 5 are presented in table 3.

Table 3. Parameters of the fast decay (value a) and slow decay (value b) component fit.

Parameter	**k1**	**k2**	**k3**	**K**	**K'**	**E**	**D**
value (a)	3.8e4 s^{-1}	5e5 s^{-1}	1e9 s^{-1}	5e6 s^{-1}	2e8 s^{-1}	150 meV	25 meV
Parameter	**k1**	**K**	**K1x**	**K2x**	**E1x**	**E2x**	
value (b)	1.9e3 s^{-1}	5e2 s^{-1}	5e8 s^{-1}	5e7 s^{-1}	5 meV	180 meV	

Slow decay points to deexcitation of one level with two quenching channels (see figure 6b). The nonradiative decay rate can be described by a simple barrier transition model:

$$k_{ix} = K_{ix}exp\left(-\frac{E_{ix}}{k_bT}\right),\ i = 1,2 \tag{4}$$

with K_{ix} being a frequency factor and E_{ix} is the height of the barrier. Parameters of the fit of the slow PL decay from figure 6b are presented in table 3.

Though the exact nature of the emission centres responsible for luminescence characteristics is not entirely clear, due to the excitation well below the band gap the luminescence process must be related to a defect and/or deeply trapped exciton. One may even consider that both emissions, i.e. that of the defect energy levels and that of the exciton contribute to the observed complex luminescence bands. Similarity of the radioluminescence band shape in all the samples suggest that the luminescence centres are independent on the chemical composition and doping of the samples, i.e. related to defects which are possibly specific for the growth technology used.

4. Conclusions

All the studied samples of single crystal borates show an emission band in UV region peaking around 310-320 nm under X-ray excitation, consistent with [3]. The intensity of samples 1, 2, 4 and 6 exceeds that of BGO standard scintillator. However, light yield measurements under alpha particle excitation of two most intensely emitting samples no. 1 and 6 provide the values which are approximately an order of magnitude lower compared to that of BGO measured in the same conditions. It points to intense slow decay components in the scintillation response. This is supported by strongly increasing LY values with increasing shaping time and by the presence of slow decay components in photoluminescence decays down to tens of microseconds.

The study of temperature dependence of the PL spectra and decays of KBBF sample no. 2 suggests that the emission band peaking at about 350 nm is composed of at least two subbands. Given the subband excitation the defect and/or deeply trapped exciton are proposed as its origin, but further study is necessary.

Acknowledgments

Financial support of Czech TACR TA01011017 project and the National Natural Science Foundation of China (Grant No. 6113800451402316) is gratefully acknowledged.

References

[1] Lin Z, Wang Z, Chen C T, Chen S K and Lee M H 2003 Mechanism for linear and nonlinear optical effects in $KBe_2BO_3F_2$ (KBBF) crystal *Chem. Phys. Lett.* **367** 523
[2] Chen C T, Wang G L, Wang X Y, Xu Z Y 2009 *Appl. Phys. B* **97** 9-25
[3] Ogorodnikov I N, Poryvay N E, Sedunova I N, Tomachev A V, Yavetskiy R P 2011 Thermally stimulated recombination processes and luminescence in Li_6(Y,Gd,Eu)$(BO_3)_3$ crystals *Physics of the Solid State* **53** 263-270
[4] Gao J H, Li R K, 2008 Potassium rich rare earth (RE) borates $K_3RE(BO_3)_2$ *Sol. St. Sci.* **10** 26-30
[5] Engels R, Reinartz R, Schelten J Czirr B 2000 Thermal neutron netection with the lithium borate scintillator *Nucl. Sci. IEEE Transac.* **47** 948-951
[6] van Eijk C.W.E 2004 Inorganic scintillators for thermal neutron detection *Rad. Meas.* **38** 337-342
[7] Singh A K, Tyagi M, Singh S G, Desai D G, Sen S, Nayak B K, Urffer M, Melcher C L, Gadkari C S 2013 Cerium doped Lithium Gadolinium Borate: A neutron Scintillator *Proceedings of the DAE Symp. on Nucl. Phys.* **58** 960
[8] Engels R, Kemmerling G, Rongen H, Schelten J, Cooper R 2002 Comparison of neutron scintillation detectors with a ^{3}He proportional counter for the spallation neutron source (SNS) *IEEE Trans. on Nucl. Sci.* **49** 923-925
[9] Ogorodnikov I N, Pustovarov V A, Yakovlev S A, Isaenko L I and Zhurkov S A 2012 Luminescence and electronic excitations in $KBe_2BO_3F_2$ crystals *ISSN 1063-7834 Phys. of the Solid State* **54**4 735–740
[10] Wang X, Yan X, Luo S and Chen Ch 2011 Flux growth of large KBBF crystals by localized spontaneous nucleation *J. of Cryst. Growth* **318** 610-612
[11] Zhou H, He X, Zhou W, Hu Z 2011 Hydrothermal growth of KBBF crystals from KOH solution *Journal of Crystal Growth* **318** 613-617
[12] Liu L, Zhou H, He X, Zhang X 2012 Hydrothermal growth and optical properties of RbBe2BO3F2 crystals *Journal of Crystal Growth* **348** 60-64
[13] D'Ambrosio C, Leutz H 2003 *Nucl. Inst. Meth. A* **501** 463-498
[14] Babin V, Gorbenko V, Krasnikov A, Makhov A, Mihokova E, Nikl M, Zazubovich S, Zorenko Y 2012 *Phys. Status Solidi* B **249** 5 1039–1045

IOP Conf. Series: Materials Science and Engineering **80** (2015) 012016 doi:10.1088/1757-899X/80/1/012016

Diameter-dependent annealing kinetics of X-ray-induced defects in single-walled carbon nanotubes

T Murakami[1], M Matsuda[1], S Isozaki[1], K Kisoda[2] and C Itoh[1]
[1]Department of Materials Science and Chemistry, Wakayama University, Wakayama 640-8510, Japan
[2]Department of Physics, Wakayama University, Wakayama 640-8510, Japan

E-mail: murakami@sys.wakayama-u.ac.jp

Abstract. Diameter-dependent annealing effects of X-ray-induced defects in single-walled carbon nanotubes (SWNTs) were studied by Raman scattering spectroscopy. We found that X-ray-induced defects were formed in thin SWNTs at higher density than that in thick SWNTs. The X-ray-induced defects, distant pairs of interstitials and vacancies (I-V), were eliminated by thermal annealing. The recovery temperature of the X-ray-induced defects in the thin SWNTs were higher than that of the thick ones, indicating the thermal stability of the defects in thinner SWNTs are higher. In order to simulate the annealing behaviours of the X-ray-induced defects in SWNTs with diameters of ~0.9 and ~1.4 nm, we suggested the competitive reactions of diffusion of interstitials, formation of close I-V pairs and their recombination. The simulated results showed that the reaction-rate constant of elimination of close I-V pairs is dependent on tube diameter, which is presumably derived from nano-structure effects of SWNTs.

1. Introduction

Single-walled carbon nanotubes (SWNTs) have characteristic electric properties originated from quasi-one-dimensional structure.[1] Their electronic property strongly depends on chirality, which describes "how to roll up" the 2D graphite sheet to make the SWNT. Because of this strong correlation between the electronic structure and the geometric structure, structural modification is one of the most feasible methods for controlling their electronic properties. However their chemical stability and mechanical toughness impede their structural modification. One of the feasible methods for modifying SWNT's structure is to utilize the defect formation induced by ionizing radiation.[2-4] Hirahara *et al* showed that the defect migration changed the chirality of a tube under the simultaneous application of tensile stress and heat by transmission electron microscope.[5] We also reported that X-ray irradiation formed distant interstitial-vacancy pairs in SWNTs, and the accumulation of the defects eventually resulted in modification of their structure.[6,7] This phenomenon is potentially available to control the SWNT's structure. In order to develop the structural modification method based on the irradiation effect, we should elucidate characteristics of the irradiation-induced defect in SWNTs. Since the SWNT samples used in our previous study [6] had wide diameter distribution, the observed results were the ensemble average over different tubes. The wide diameter distribution may veil the dependence of the defect process on the tube diameter.

In the present report, we show the results of the study on thermal annealing kinetics of the X-ray-induced defects in SWNTs with narrow diameter distributions, and discussed their thermal stability in terms of the tube diameter. We found that the X-ray irradiation induced defects at higher density in thinner tube and complete recovery of the X-ray-induced defect required higher sample temperature in thinner tube. The recovery process of the defects was hardly explained by a simple diffusion-limited recombination, and thus we made a simulation with considering competitive-recovery reactions composed of diffusion-limited recombination and the reaction involving formation of intermediate state, like close I-V pair. The results showed that the reaction-rate constant of elimination of close I-V pairs was dependent on tube diameter. The diameter dependence of the annealing effect is presumably derived from nano-structure effects of SWNTs.

2. Experimental procedure

SWNT samples synthesized by arc-discharge technique and CoMoCAT CVD method were purchased from Aldrich Co., Ltd. Their mean diameters were ~1.4 and 0.9 nm, evaluated by Raman measurements. The purchased SWNTs were dispersed in aqueous solutions of 1-wt% sodium dodecyl sulfate using a tip-type sonicator. The dispersed solutions were centrifuged, and then the supernatant filtered using a mixed cellulose ester membrane. After the membrane was removed by acetone, the SWNT films were transferred on silicon substrates. The SWNT films were irradiated with X-ray (1.24 keV) in vacuum. The irradiated samples were taken in the air and subsequently annealed at the given temperatures for 30 min. The annealing temperature was raised sequentially in each annealing from 100 to 700 °C by the step of 50-100 °C. The irradiated and annealed SWNT samples were characterized by Raman scattering spectroscopy at room temperature using a probe laser of 532 nm.

3. Results and Discussion

Figure 1 shows the Raman spectra of the pristine, irradiated, and annealed (600 °C) SWNTs synthesized by (a) arc-discharge technique and (b) CoMoCAT CVD method. Tangential vibration modes of ~1600 cm^{-1} (G band) and defect-induced mode of ~1350 cm^{-1} (D band) are clearly observed for both samples. The Raman intensities were normalized with respect to the G band. Insets in Fig. 1(a) and (b) show radial breathing modes (RBMs) of pristine samples. The diameter of SWNT (d) can be estimated from RBM frequency (ω_{RBM}).[8] We evaluated d~1.4 and ~0.9 nm from the observed peaks of ω_{RBM}~170 and ~270 cm^{-1}, respectively. Hereafter, we describe the SWNT of d~1.4 nm as the thick tube and describe the SWNT of d~0.9 nm as the thin tube.

X-ray irradiation enhanced the intensity of the D band, indicating that the irradiation induced defects in SWNTs. We found that the D band enhancement of the thin tube was much larger than that of the thick tube. This result shows the thicker tube is more tolerant to the radiation. The increases in the D band intensity of both irradiated samples were reduced by annealing at 600 °C. This recovery was ascribed to the recombination of vacancy and interstitial pair. While the net increase of the D band in thick tube recovered by the annealing, that of the thin tube did not fully recover even at the maximum annealing temperature in the present study. The slight changes of the Raman spectra are possibly related to the modification of tube structure by the X-ray irradiation and subsequent annealing.

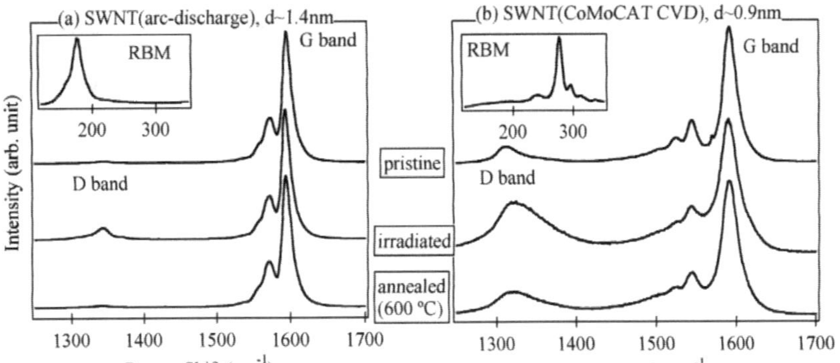

Figure 1. Raman spectra of pristine, irradiated, and annealed SWNTs of (a) d~1.4 nm and (b) d~0.9 nm. Insets show RBMs of pristine samples.

In Fig. 2, we plotted the ratio of the changes in the net increase of D band by the X-ray irradiation to G band (ΔD/G) as a function of the annealing temperature. The annealing behaviors of each tube are remarkably different in two aspects: the completing temperature of the annealing and the magnitude of a hump in the annealing profile around 300 °C. While ΔD/G of thick tube almost faded out near 400 °C, that of thin tube showed decrease up to 600 °C. This results shows that the defects formed in the thin tube are more stable than those in the thick tube. The annealing profiles of each tube showed a broad hump ranging from 200 °C to 400 °C. The magnitude of the hump is larger in the thin tube than the thick tube. The results suggest that the defect recovery process is not simple and depends on the tube diameter.

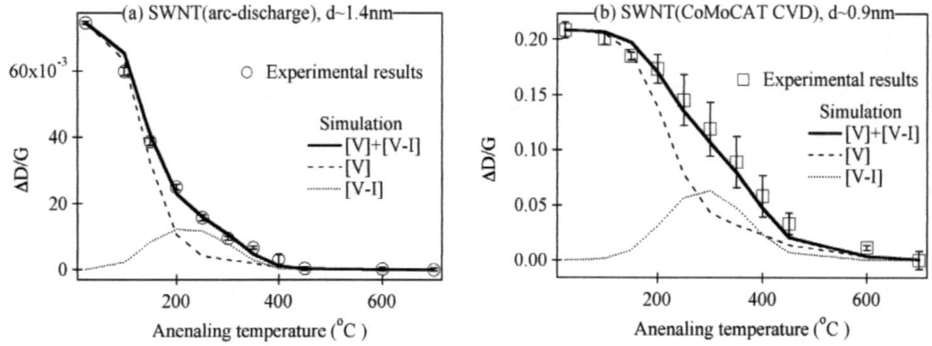

Figure 2. ΔD/G of SWNTs with diameter of (a) ~1.4 nm and ~0.9 nm, depending on annealing temperature. Solid, dashed, and dotted lines show the simulated results.

The X-ray-induced defect in SWNT has a form of distant vacancy-interstitial pair.[6] Hence, the recombination of the defect is controlled by diffusion process. Our previous study using SWNT samples with wide diameter distribution showed the defect-recovery process was successfully simulated by a 2nd-order reaction model.[6] The simple simulation used in our previous study hardly explains the hump found in the present study. Since the samples used in our previous study were mixture of SWNTs with various diameters, the observed signals were ensemble average of the signals of each SWNT. Hence, the detail annealing features depending on the tube diameter were probably averaged out.

In order to explain the annealing profiles obtained in the present study, we presume that intermediate states are formed prior to the interstitial-vacancy (I-V) recombination. One of the most probable candidates for the intermediate state is close I-V pair. When the close I-V pair instantaneously recombines after the formation, the pair formation process dominates the recombination reaction. However, the recombination rate is much slower than the pair formation rate, the recombination of the close pair is the rate-determining process. In SWNT, it is possible that the recombination rate of the close pair depends on the configuration of the pair. Thus, we presume that two types of close I-V pairs are formed as the intermediate states: one has a recombination rate much larger than the formation rate and the other has a recombination rate smaller than the formation rate. The reaction schemes are described below:

$$V + I \underset{k_2}{\overset{k_1}{\rightleftarrows}} (I \cdot V)_A \xrightarrow{k_3} P \tag{1}$$

$$V + I \underset{k_5}{\overset{k_4}{\rightleftarrows}} (I \cdot V)_B \xrightarrow{k_6} P \tag{2}$$

Here, V, I, and P denote a vacancy, an interstitial, and a perfect lattice in an SWNT, respectively. Two close I-V pairs are designated as $(I\text{-}V)_A$ and $(I\text{-}V)_B$. k_i (i=1,2,3,4,5,6) are the reaction rate constants. Thermal activation type temperature dependence of the reaction rate constants, $k_i = k_{0i}\exp(-E_{ai}/k_B T)$, is considered. k_{0i}, E_{ai}, and k_B are pre-exponential factors, activation energies, and Boltzmann constant, respectively. Based on our presumption, we set the conditions $k_1 << k_3$ and $k_6 << k_4$. The former condition provides that the formation of $(I\text{-}V)_A$ is the rate-determining process of the reaction scheme (1), and the later provides that the recombination of $(I\text{-}V)_B$ is the rate-determining process of the scheme (2). Consequently, the reaction velocities are described as bellow differential equations.

$$d[V]/_{dt} = -(k_1 + k_4)[V][I] + k_5[I \cdot V] = -(k_1 + k_4)[V]^2 + k_5[I \cdot V] \quad (3)$$

$$d[I \cdot V]/_{dt} = k_4[V][I] - (k_5 + k_6)[I \cdot V] = k_4[V]^2 - (k_5 + k_6)[I \cdot V] \quad (4)$$

[V], [I], and [I-V] are the densities of the vacancy, the interstitial, and the close I-V pair in SWNTs, respectively. Because the X-ray-induced defects composed of vacancy and interstitial, [V] is equal to [I]. ΔD/G are generally proportional to the defect density.[9] In the present simulation, we suppose that all of the vacancies contributes ΔD/G equally. Hence, ΔD/G is proportional to the sum of [V] and [I-V]. We optimized parameters of k_1, k_4, k_5, k_6 and simulated using numerical analytical approach. The simulated results are shown as the solid curves in Fig. 2. The result shown in the figure is one of the results successfully reproducing the experimental points. The dashed and dotted curves show the contributions of [V] and [I-V], respectively. While [V] is dominated by the interstitial diffusion, [V-I] is governed by the recombination of close I-V pair. In order to simulate the annealing profile of the thin tube, we suppose the larger contribution of close I-V pair in the thin tube than that in the thick tube. Presumably, k_6 depends on the tube diameter and the dependence is derived from nano-structure effects of SWNTs. The local configuration of I-V pair would be affected by curvature of the wall. This curvature effect is stronger in the thin tube than thick tube.

On the basis of these results, we suggest that the recombination of distant I-V pair involves formation of intermediate state, like close I-V pair. In the present study, we cannot determine the best parameters of k_{0i} and E_{ai}, because many parameters were used in our simulation. In order to reveal the tube-diameter dependence of these parameters, further approaches are required such as an isothermal annealing experiment, direct observation of the defects using TEM or atomic force microscope, and theoretical calculations.

4. Conclusion

Diameter-dependent annealing kinetics of the X-ray-induced defects in SWNTs was studied by Raman scattering spectroscopy. The thermal stability of the X-ray-induced defects in thin SWNTs (d~0.9 nm) were higher than those of thick SWNTs (d~1.4 nm). The annealing profiles of these tubes were explained by concerning the formation of intermediate stats, like close I-V pairs. The yield of the intermediate state depends on tube diameter; the I-V pairs' elimination in thick SWNTs has the higher reaction-rate constant than that in thin SWNTs. In order to clarify the recovery reaction of the X-ray-induced defects in detail, further experimental and theoretical studies are required.

References

[1] Rotkin S V and Sunramoney S 2005 *Applied Physics of Carbon Nanotubes: Fundamentals of Theory, Optics and Transport Devices* (Springer: Berlin) p 227
[2] Pomoell J A V, Krasheninnikov A V, Nordlund K and Keinonen J 2004 *J. Appl. Phys.* **96** 2864
[3] Suzuki S and Kobayashi Y 2007 *J. Phys. Chem.* C **111** 4524
[4] Itoh C, Uotome K, Kisoda K, Murakami T and Harima H 2008 *Nucl. Instrum. Methods Phys. Res., Sect.* B **266** 2772
[5] Hirahara K, Inose K and Nakayama Y 2010 *Appl. Phys. Lett.* **97** 051905
[6] Murakami T, Yamamoto Y, Kisoda K and Itoh C 2013 *J. Appl. Phys.* **114** 114311
[7] Murakami T, Yamamoto Y, Matsuda M, Kisoda K and Itoh C 2014 *Jpn. J. Appl. Phys.* **53** 02BD11
[8] Jorio A, Saito R, Hafner J H, Lieber C M, Hunter M, McClure T, Dresselhaus G and Dresselhaus M S 2001 *Phys. Rev. Lett.* **86** 1118
[9] Cançado L G, Jorio A, Ferreira E H M, Stavale F, Achete C A, Capaz R B, Moutinho M V O, Lombardo A, Kulmala T S and Ferrari A C 2011 *Nano Lett.* **11** 3190

IOP Conf. Series: Materials Science and Engineering **80** (2015) 012017 doi:10.1088/1757-899X/80/1/012017

Lanthanum – Gallium Tantalate Crystals: Surface processes and their effect on electrophysical properties

N S Kozlova[1], O A Buzanov[2], A P Kozlova[1, 1], and I M Anfimov[1]

[1] National University of Science and Technology "MISIS", Moscow, Russia
[2] Company Fomos-Materials, Moscow, Russia

E-mail: anna_kozlova_2009@mail.ru

Abstract. The behaviour of short - circuit currents in specimens of the polar cuts of lanthanum-gallium tantalate crystals with identical deposited electrodes before and after preliminary polarization is investigated. The temperature dependences of short - circuit currents in the temperature range 200-600 °C were measured. These currents are caused by the electrochemical processes on opposite polar cuts surfaces taking place at the crystal /electrode boundary.

1. Introduction

Currently, the investigation and application of lanthanum-gallium tantalate crystals ($La_3Ga_{5,5}Ta_{0,5}O_{14}$, langatate, LGT) is an intensively developing direction of the research in piezoelectronics (in particular, in the production of high-temperature devices based on the direct piezoelectric effect [1, 2, 3]).

One of the most important requirements for crystals used in piezoelectronics is their physical properties stability determined by the nonequilibrium state of the crystal and the kinetics of the process of charge transfer.

The nature of the nonequilibrium state of crystals and, consequently, the possibility of occurrence of processes leading to changes of their state may be caused by the properties of the crystal itself, by its production history and operation conditions. It is well known that there are deviations from the equilibrium caused by external effects, for example, degradation of crystalline elements under the effect of electric fields. Less evident are processes and their consequences determining the variation of crystals in the absence of external effects. This process includes the short - circuit current (SCC) detected in langatate [3, 4].

The technology of high-temperature devices based on langatate crystals includes the coating of the surfaces of their polar cuts with a thin layer of a conducting material. It is known [4] that the deposition of metal electrodes on the surface of the polar cuts of crystals leads to their aging and degradation due to the occurrence of electrochemical processes. In this context, one of the most important demands in designing high-temperature sensors based on lanthanum–gallium tantalate is the proper choice of the optimum material of the conducting coating.

2. Experiment results

We have shown that the short - circuit currents in LGT grown in the atmospheres of argon and argon with oxygen on level of $10^{-14}\div10^{-8}$ A are detected always even if symmetric (identical) electrodes are placed on the polar cuts of the crystal. The short - circuit currents were observed on all specimens without preliminary polarization with symmetric electrodes from different materials (gold, gold with a titanium sublayer, iridium, silver with a chromium sublayer) [3, 5].

[1] Author to whom any correspondence should be addressed

Short - circuit currents can be induced in solid insulators using the following techniques:

- symmetric (identical) electrodes are deposited on the crystal surface, and the samples are preliminarily stimulated using thermo-, photo-, electric, X-ray, or piezoelectric effects or polarization at high temperatures.
- asymmetric electrodes (E_1, E_2) are deposited on the insulator (I) surface according to the scheme $E_1/I/E_2$. In this case, the short - circuit currents are observed without preliminary polarization. This is due to the difference in the electrochemical processes occurring during the contact of the insulator surface with different metal electrodes, as well as due to the different activity of the electrode coatings.

If the crystal between two electrodes is examined as a galvanic cell, then for a cell with the solid electrode of the type $E/E_1/I/E_2/E$, where I is a solid electrolyte, E is the material conducting leads, the value of electromotive force (emf) is written in the form:

$$\varepsilon = \varphi_g^{(E_2 I)} - \varphi_g^{(E_1 I)} + \varphi_g^{(E_2 E_1)}$$

where φ_g is the Galvani potential at the electrode/ electrolyte phase boundary, ε is the electrode potential at the i-th electrode.

On the basis of the experimental investigation of near-electrode surface of a polar crystals, e.g. α-$LiIO_3$ [6] and LGT [3, 5], it was determined that the origin of the short - circuit currents observed in the polar cuts without application of an external electric field or any another preliminary polarization was due to generation of an intrinsic emf (0.02÷0.04 V) as a result of the electrochemical processes on the sample surface coming in contact with the electrodes.

The object examined in this work can be represented as an electrochemical cell with symmetric electrodes $E/E_1/pr_1/I/pr_2/E_1/E$, where I is a solid electrolyte, E is the material conducting leads, E_1 is the material of the electrode coatings, pr1 and pr_2 are the products of electrochemical reactions. A standard electrochemical cell with solid electrolytes and symmetric electrodes should have zero emf. However, when polar cuts of the polar crystals are used as a solid electrolyte, an emf is generated due to the different electrochemical processes on the opposite surfaces of the polar cuts as a result of their different activity.

Previously in [3, 5] it was shown the SCC changes 4 to 6 orders of magnitude depending on electrode material. Therefore, in this study, we performed experiments with langatate crystals with the identical electrodes of Ta before and after preliminary polarization. This choice of electrode material is related to the fact that tantalum is part of the single crystal composition. The crystals were produced by the "Fomos-Materials" company and were grown in iridium crucibles in a mixture of argon and oxygen ($Ar+(2\%)O_2$). The samples were made in form of plates of polar cut polished on both sides with magnetron sputtered electrodes. The thickness of the deposited electrodes was approximately 200 nm. The electrical parameters were measured on a "Keithley" 6517 A electrometer. SCC measurements were carried out in the range 20 – 600 °C, the heating rate was constant (2 °C/min). Charge state of each surface was determined by the piezoresponse technique. In the experiment, the samples were always installed in a crystal holder considering the surface charge sign.

SCC dependence of LGT samples with Ta electrodes is shown in Figure 1a. We observed a well-defined peak, which can be divided into two different. This fact shows good stability of the electrode-dielectric-electrode system and the absence of serious competing processes that could contribute in the short - circuit currents.

The preliminary polarization of specimens was carried out at a field of 50 V/mm while heating up to 500 °C, then the sample was kept for one hour at this temperature and rapidly cooled. The preliminary stage is necessary to detect the contribution in SCC of greater number of defects than without preliminary polarization. This effect is related to the fact that during the voltage supply and heating the following processes occur in the sample: processes of space-charge generation and their polarizations enhanced by heating and thermally stimulated conductivity. The short - circuit currents are recorded after preliminary polarization and this dependence is shown in Figure 1b.

After the preliminary polarization of the sample value of the current amplitude decreased by 2 orders of magnitude (10^{-8} to 10^{-10} A). The equilibrium conductivity near the electrodes leads to the accumulation of the volumetric charges of opposite sign under the application of an external electric field. These charges screen electric field generated in polar crystal.

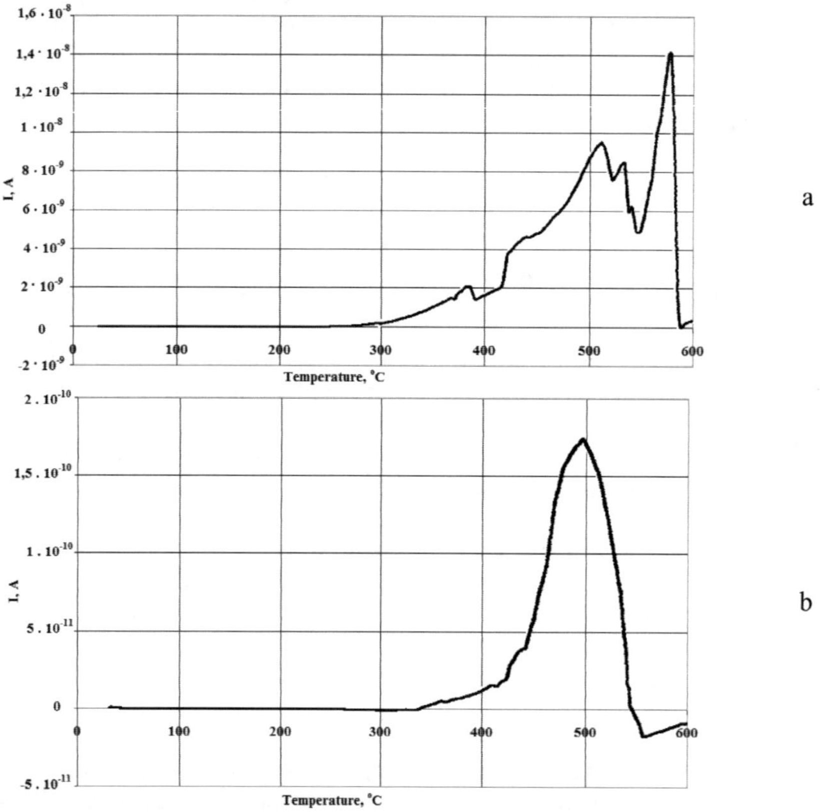

Figure 1. SCC dependence of LGT samples with Ta electrodes before (a) and after preliminary polarization (b).

The method of mathematical processing of SCC experimental data– the method of initial rise – is described in [4]. This method is based on the results of theoretical investigation of thermo-stimulated currents. It was shown that the initial phase of thermally stimulated current is linearized in the coordinates of the Arrhenius ln I(1/kT) regardless of the mechanism and order of kinetics of non-isothermal relaxation of the charge of (figure 2).

The slope of a line, resulting in the construction of the initial phase of the peak thermo-stimulated current in the coordinates of the Arrhenius, allows calculation of the activation energy W at $T \approx T_0$ [4]:

$$-W \approx d\ln I(T) \big/ d(\frac{1}{kT}) \qquad (1)$$

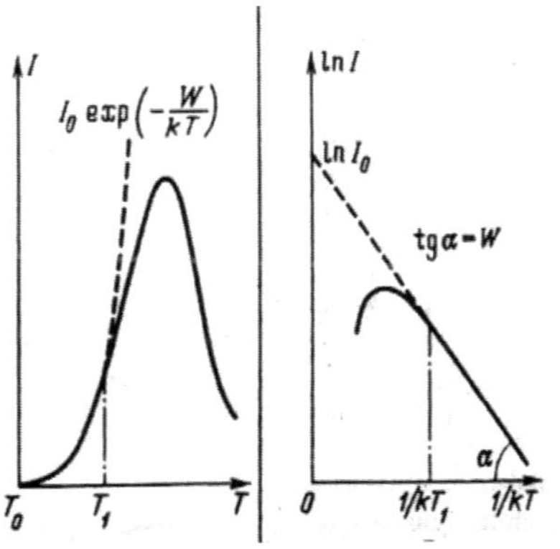

Figure 2. The methods of initial rise.

The advantage of the described method is its independence from the order of the kinetics of the relaxation process.

The values of activation energies show the dominance of one process over the other at a certain temperature. This allows t separation of the contribution of the near-electrode processes in the release of electrically-active defects analyzing the temperature dependence of the SCC. The values of activation energy for the sample before and after the preliminary polarization are shown in table 1.

Table 1. Characteristics of the SCC peaks

Temperature (°C)	Peak current (A)	Activation energy by the methods of initial rise (eV)
	Before preliminary polarization	
575	$1{,}5 \cdot 10^{-8}$	1,0
	After preliminary polarization	
495	$1{,}8 \cdot 10^{-10}$	2,2

References

[1] Shimamura K, Takeda H, Kohno T 1996, *J. Cryst. Growth* **163** 388

[2] Takeda H, Tanaka S, Izukawa J 2005 *in Proc. 2005 Ultrasonic Symp. IEEE.* **1** 560.

[3] Buzanov O A, Zabelina E V, Kozlova N S, Sagalova T B 2008 *Crystallography Reports* **53** 853–857

[4] Gorokhovatskiy Yu A and Bordovskiy G A 1991 *Thermally Activational Current Spectroscopy of High-Resistance Semiconductors and Dielectrics* (Nauka, Moscow) p 248 [in Russian]

[5] Kozlova N S, Kozlova A P, Anfimov I M, Kiselev D A, Bykov A S 2014 Lanthanum-gallium tantalate crystals and their electrophysical characterization *Journal of Nano- and Electronic Physics* **6** 03034

[6] Blistanov A A, Kozlova N S and Geras'kin V V 1997 The Phenomenon of electrochemical self-decomposition in polar dielectrics *Ferroelectrics* **198** 61-66

IOP Conf. Series: Materials Science and Engineering **80** (2015) 012018 doi:10.1088/1757-899X/80/1/012018

Primary and aggregate color centers in proton irradiated LiF crystals and thin films for luminescent solid state detectors

M Piccinini, F Ambrosini, A Ampollini, F Bonfigli, S Libera, L Picardi, C Ronsivalle, M A Vincenti and R M Montereali

ENEA C.R. Frascati, Via E. Fermi 45, 00044 Frascati (Rome), Italy

E-mail: massimo.piccinini@enea.it

Abstract. Proton beams of 3 MeV energy, produced by the injector of a linear accelerator for proton therapy, were used to irradiate at room temperature lithium fluoride crystals and polycrystalline thin films grown by thermal evaporation. The irradiation fluence range was 10^{11}-10^{15} protons/cm^2. The proton irradiation induced the stable formation of primary and aggregate color centers. Their formation was investigated by optical absorption and photoluminescence spectroscopy. The F_2 and F_3^+ photoluminescence intensities, carefully measured in LiF crystals and thin films, show linear behaviours up to different maximum values of the irradiation fluence, after which a quenching is observed, depending on the nature of the samples (crystals and films). The Principal Component Analysis, applied to the absorption spectra of colored crystals, allowed to clearly identify the formation of more complex aggregate defects in samples irradiated at highest fluences.

1. Introduction

Color centers (CCs) in lithium fluoride (LiF) crystals and thin films found application in optically pumped solid-state lasers [1] and miniaturised light-emitting devices [2], due to the high efficiency of their photoluminescence (PL) process. Recently, LiF crystals and thin films have been also proposed as novel solid-state soft x-ray and neutron imaging detectors [3,4], based on the optical reading of the visible PL of the radiation-induced F_2 and F_3^+ (two electrons bound to two and three close anion vacancies, respectively) CCs. These laser-active defects possess almost overlapping absorption bands peaked around 450 nm, called M band [5], and, under light excitation in this spectral range, they emit broad PL bands peaked at 678 nm and 541 nm for F_2 and F_3^+ CCs, respectively.

On the other hand, ion beams of different energies are widely investigated for applications ranging from material modifications [6,7] to radiobiology and radiotherapy. With the aim to use PL of point defects in LiF for proton beam monitoring and dosimetry, we recently started the investigation of the optical properties of stable CCs induced by proton beams in LiF crystals and thin films [8].

Proton beams of 3 MeV energy, produced by a linear accelerator (LINAC) used as the injector of a proton therapy LINAC, under development at ENEA Frascati [9], were used to irradiate at room temperature (RT) LiF crystals and polycrystalline thin films grown by thermal evaporation [2]. The irradiation fluence range was 10^{11}-10^{15} protons/cm^2. Proton irradiation induced the stable formation of primary (F) and aggregate (mainly F_2 and F_3^+) CCs. Their formation was investigated by optical absorption and photoluminescence spectroscopy and the Principal Component Analysis (PCA) was applied to the absorption spectra of colored crystals.

2. Materials and Methods

Exposed samples were 10x10x1 mm^3 LiF crystals polished on both faces, commercially available (MacroOptica Ltd.), and polycrystalline LiF thin films, about 1 μm thick, grown by thermal evaporation on glass substrates [10] kept at a constant temperature of 300 °C during the deposition process, performed in a vacuum chamber at a pressure below 1 mPa, at the Solid State Lasers Laboratory in ENEA Frascati. The starting material consists of LiF microcrystalline powder (Merck Suprapur, 99.99% pure), heated at about 800°C in a water-cooled tantalum crucible. The evaporation rate, monitored in situ by an INFICON quartz oscillator, was automatically controlled at a fixed value of 1 nm/s during the growth.

Proton beams of 3 MeV energy were produced by the PL7 model LINAC by ACCSYS-HITACHI. At the output of the machine beamline a 50 μm thick kapton window was placed, which reduced the impinging protons energy to 2.23 MeV. LiF samples were irradiated in air at RT.

During proton irradiation, LiF crystals and films were fixed on an aluminium mask with a 3 mm pin-hole, in order to irradiate them on circular spots with the highest and most uniform transversal intensity distribution of the proton beam. The average beam current was 1 μA in 60 μs-long pulses at a repetition frequency of 50 Hz. The irradiation fluence covered the range of 10^{11}-10^{15} protons/cm^2 by varying the total number of pulses delivered to different LiF samples.

Optical absorption spectra were acquired at RT on the irradiated areas of LiF crystals by a Perkin-Elmer Lambda 950 spectrophotometer with 1 nm resolution. The PL spectra of LiF crystals and films were measured at RT by pumping in a continuous wave regime with the 457.9 nm line of an Argon laser, which allows to simultaneously excite the green and red emissions of F_3^+ and F_2 CCs [11]. The PL signal was spectrally filtered by a monochromator and acquired by means of a photomultiplier with lock-in technique. The PL spectra were corrected for the instrumental calibration.

3. Results and Discussion

Figure 1 shows the PL spectra of LiF crystals irradiated by 3 MeV protons at seven fluence values in the interval from 1.5×10^{12} to 6.4×10^{14} protons/cm^2. They consist of two broad emission bands peaked around 540 nm and 670 nm, ascribed to F_3^+ and F_2 centers, respectively [11]. The emission band intensities are proportional to the fluence up to values of the order of 10^{14} protons/cm^2, after which a quenching starts to be observed. The PL spectra of LiF films irradiated in the same experimental conditions (not shown) exhibit the same spectral features of LiF crystals.

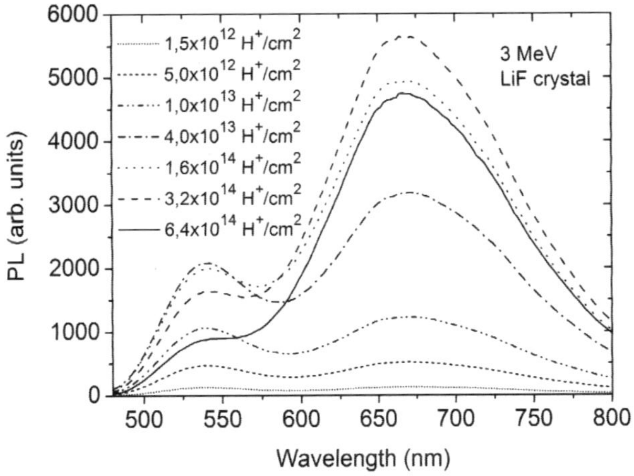

Figure 1. Photoluminescence spectra of LiF crystals irradiated by a 3 MeV proton beam at seven fluence values, measured at room temperature under laser pumping at 457.9 nm.

In order to measure the spectral contributions of F_2 and F_3^+ defects to the total visible PL intensity and to investigate their behaviour with fluence, after converting the wavelength from nm to eV, the rescaled PL spectra were fitted as the sum of two Gaussian bands. The areas under the bands, associated with the contribution to the total PL of the F_2 and F_3^+ CCs, were obtained for every spectrum and plotted against the proton irradiation fluence. The results are shown in figure 2.

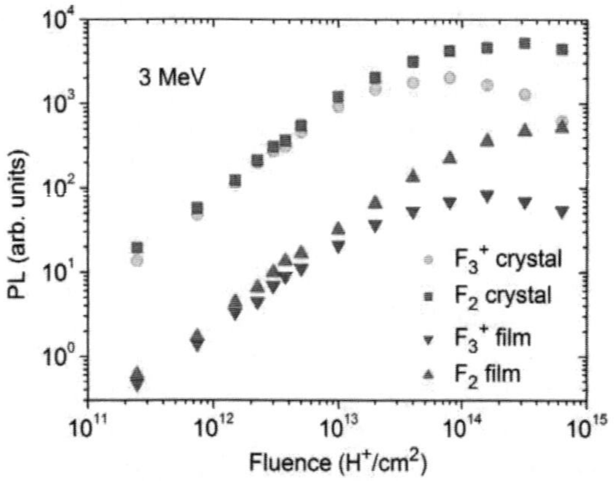

Figure 2. Visible photoluminescence signal of F_2 and F_3^+ color centers as a function of the 3 MeV proton fluence in colored LiF crystals and 1 μm thick films grown on a glass substrate.

According to figure 2, the PL intensity of both F_2 and F_3^+ CCs in crystals is more than one order of magnitude higher than in LiF films. In LiF crystals, radiation induced defects are produced inside a volume whose depth is comparable with protons implantation depth (45 μm at 2.23 MeV, according to simulations performed by SRIM [12]), within which they lose all their energy [13]. On the contrary, in LiF films the collected PL is emitted by CCs located in a LiF layer of thickness 1 μm, where only a small fraction of the total proton energy is lost, the rest being released in the glass substrate.

The PL intensity behaviour with fluence of F_2 CCs shows a linear increase up to $2.0x10^{13}$ protons/cm^2, while at higher fluences saturation effects start taking place, resulting into a PL sub-linear behaviour. Also in LiF films the PL signal is proportional to the fluence values, but the linear optical response reaches the fluence value of $8.0x10^{13}$ protons/cm^2, then turning into sub-linear. Concerning the PL intensity behaviour of F_3^+ CCs with fluence, it shows a linear increase up to 10^{13} protons/cm^2 in both LiF crystals and films. The PL intensity difference between F_2 and F_3^+ CCs in both crystals and films is almost constant up to a fluence of 10^{13} protons/cm^2, but with further increasing fluence, such a difference starts increasing. In particular, in the fluence interval from 10^{13} to 10^{14} protons/cm^2 the PL intensity of F_3^+ CCs increases sub-linearly with fluence, but starts decreasing from fluence values of 10^{14} protons/cm^2 in LiF crystals and $2.0x10^{14}$ protons/cm^2 in LiF films.

The F_2 and F_3^+ PL quenching effects observed at higher fluence values could be explained by thoroughly examining the absorption spectra of irradiated LiF crystals.

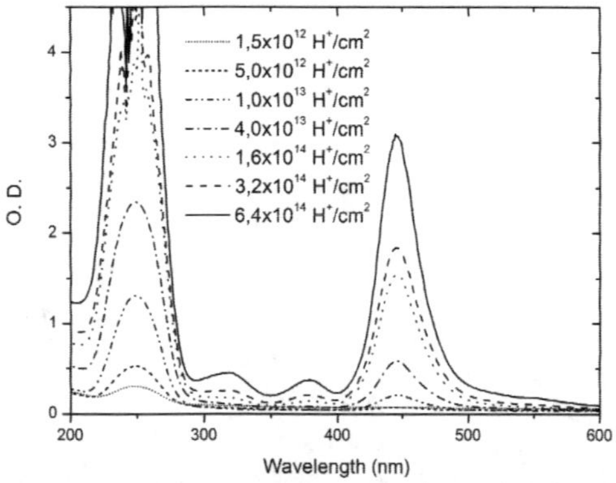

Figure 3. Optical absorption spectra of LiF crystals irradiated by a 3 MeV proton beam at seven fluence values, measured at room temperature.

Figure 3 shows the optical absorption spectra of LiF crystals irradiated by 3 MeV protons at seven fluence values in the interval from $1.5x10^{12}$ to $6.4x10^{14}$ protons/cm^2. The main contributions are due to absorption bands peaked around 250 nm and 440 nm, ascribed to primary F centers and to the superposition of the F_3^+ and F_2 centers absorption bands (M band), respectively [5,11]. For crystals irradiated with fluences higher than 10^{14} protons/cm^2, other bands could be observed. They were peaked around 320, 380 and 550 nm and are ascribed to $F_3(R_1)$, $F_3(R_2)$ and $F_4(N_2)$ more complex aggregate centers, respectively [7,14]. The last one could contribute to the observed F_3^+ PL quenching effect, causing re-absorption.

The application of PCA [15] to the second derivative of the absorption spectra in the 300-600 nm spectral range allowed us to extract more detailed information. PCA allows one to reduce the dimensionality of a data set while retaining as much as possible of its variation, hence showing data in a simplified manner.

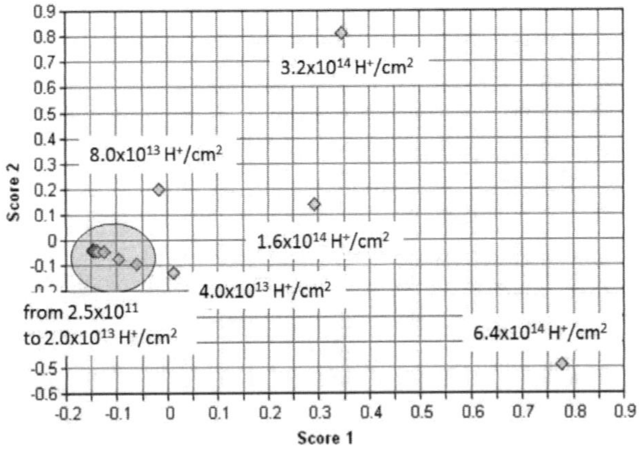

Figure 4. Principal Component Analysis score plot showing the variances associated with the first two principal components (Score 2 vs Score 1). The ellipse shows the cluster formed by dots representing the spectra of samples irradiated up to a fluence of 2.0x10^{13} protons/cm^2.

In figure 4 is reported the score plot showing the variances associated with the first two principal components, which contain most of the variation present in all spectra. Up to a fluence of 2.0x10^{13} protons/cm^2, the spectra (represented by dots) belong to the same cluster (circled), as they are very similar. When the fluence value gets higher, the spectra are not clustered, as they start to be different from each other. This result shows a clear relationship between the absorption spectra and the PL sub-linear behaviour, which started from the fluence value of 2.0x10^{13} protons/cm^2, as shown in figure 2.

The main peaks of the first two principal components (not shown) have confirmed the presence of absorption bands at 320, 380 and 550 nm, but three more bands have been found at 520, 430 and 465 nm. The first one can be ascribed to $F_4(N_1)$ CCs [14], while the other ones cause M band broadening, whose main components could be related to the presence of nanometric lithium colloids [7,16]. Their formation is due to complex phenomena related to the high proton energy release, especially at the end of the penetration path (Bragg peak) [6].

4. Conclusions

The PL intensity of 3 MeV proton-induced F_2 and F_3^+ color centers was measured in LiF crystals and thin films in a wide interval of irradiation fluence. At the lowest fluences a linear optical response of their PL signal was obtained for both kinds of samples. At the highest fluences, the F_2 and F_3^+ PL signal reaches saturation, but in LiF crystals it is achieved at lower values than in LiF films. The PCA was applied to all the optical absorption spectra of proton irradiated LiF crystals, to highlight the formation of more complex aggregate defects, among them F_3, F_4 and nanometric lithium colloids, whose complex spectral features often are superimposed to the broad absorption bands of F_2 and F_3^+ defects. PCA showed to be a very powerful tool to extract the spectral information from optical absorption spectra, where overlapping bands often coexist. It can support systematic studies of the PL response of proton beam detectors based on CCs PL in LiF crystals and thin films.

Acknowledgments

Research carried out within the TOP-IMPLART (Oncological Therapy with Protons - Intensity Modulated Proton Linear Accelerator for RadioTherapy) project, funded by Regione Lazio, Italy.

References

[1] Basiev T T, Mirov S B and Osiko V V 1988 *IEEE J. Quantum Electron.* **24** 1052

[2] Montereali R M 2002 *Handbook of Thin Film Materials* vol 3, ed H S Nalwa (San Diego: Academic Press) chapter 7 pp 399-431

[3] Baldacchini G *et al.* 2005 *Rev. Sci. Instrum.* **76** 113104_1

[4] Matsubayashi M, Faenov A, Pikuz T, Fukuda Y and Kato Y 2010 *Nucl. Instrum. Meth.* A **622** 637

[5] Nahum J and Wiegand D A 1967 *Phys. Rev.* **154** 3 817

[6] Abu-Hassan L H and Townsend P D 1986 *J. Phys. C, Solid State Phys.* **19** 99

[7] Perez A, Balanzat E and Dural J 1990 *Phys. Rev.* B **41** 7 3943

[8] Piccinini M, Ambrosini F, Ampollini A, Carpanese M, Picardi L, Ronsivalle C, Bonfigli F, Libera S, Vincenti M A and Montereali R M 2014 *Nucl. Instrum. Meth.* B **326** 72

[9] Ronsivalle C *et al.* 2011 *Eur. Phys. J. Plus* **126** 7 68

[10] Montereali R M, Baldacchini G, Martelli S and Scavarda do Carmo L C 1991 *Thin Solid Films* **196** 75

[11] Baldacchini G, De Nicola E, Montereali R M, Scacco A and Kalinov V 2000 *J. Phys. Chem. Solids* **61** 21

[12] Ziegler J F, Ziegler M D and Biersack J P 2010 *Nucl. Instrum. Meth.* B **268** 1818

[13] Piccinini M, Ambrosini F, Ampollini A, Carpanese M, Picardi L, Ronsivalle C, Bonfigli F, Libera S, Vincenti M A and Montereali R M 2014 *J. Lumin.* **156** 170

[14] Montecchi M, Nichelatti E, Mancini A and Montereali R M 1999 *J. Appl. Phys.* **86** 6 3745

[15] Jolliffe I 2005 *Encyclopedia of Statistics in Behavioral Science* vol 3, ed B S Everitt and D Howell (Chichester: John Wiley & Sons, Ltd)

[16] Mussi V, Cricenti A, Montereali R M, Nichelatti E, Scaglione S and Somma F 2002 *Phys. Chem. Chem. Phys.* **4** 2742

Optimization of Immobilization of Nanodiamonds on Graphene

A Pille[1], S Lange[1], K Utt[1], M Eltermann[1]

[1]Institute of Physics, University of Tartu, Ravila 14c, 50411, Tartu, Estonia

E-mail: annika.pille@ut.ee

Abstract. We report using simple dip-coating method to cover the surface of graphene with nanodiamonds for future optical detection of defects on graphene. Most important part of the immobilization process is the pre-functionalization of both, nanodiamond and graphene surfaces to obtain the selectiveness of the method. This work focuses on an example of using electrostatic attraction to confine nanodiamonds to graphene. Raman spectroscopy, micro-luminescence imaging and scanning electron microscopy were applied to characterize obtained samples.

1. Introduction

Graphene is a two-dimensional atomically thin semi-metallic film, where carbon atoms are arranged into sp^2 honeycomb lattice. Graphene has excellent mechanical, optical, thermal and electrical properties and finding a way to employ them easily is of most importance. There have been reports on graphene oxide- semiconductor/metal nanoparticle composites as a probable tool for water splitting, solar cells and fuel cells [1]. Learning to prepare and functionalize graphene surface with desired properties easily and cost-efficiently holds great potential for future bottom-up electronics.

Dealing with graphene means manipulating it in nano-scale, which makes the process time-consuming, expensive and prone to defect creation. Defects in graphene can manifest in many forms and at different scale. Atomic scale defects like mono- and multi-atom vacancies can seriously disturb graphene electron structure leading to reduced conductivity [2]. More sophisticated defects like dislocations, edges or Stone-Wales defects can enhance the chemical reactivity of graphene compared to perfect lattice [3]. Micro-scale defects are also relatively common and naturally occur due to stresses caused by substrate thermal expansion or mechanical manipulation (e.g. Scotch tape technique) [4]. Uneven growth thickness of graphene can often form islands of multilayer graphene in micro-scale.

In all cases a robust and easy method for defect detection and quantification is desired. As graphene sheets are usually grown in an atom thick layer, it is very difficult to distinguish between mono- and multilayer graphene solely by looking at optical or SEM micrographs. This leads to Raman spectroscopy being a commonly used tool in graphene investigations [5].

In present report a spectroscopic method for optical identification of micro-scale defects on graphene is introduced via selective marking of defective areas with luminescent nanodiamonds (ND). The selective adhesion of ND-s is achieved via zeta potential modification of both graphene or its defects and the ND-s to introduce electrostatic attraction in a liquid solvent.

Nanodiamonds with nitrogen-vacancy (N-V) centers are selected as markers because of their well-known luminescence properties. They are readily used as fluorescent markers in drug delivery and bio-imaging [6] but also as single photon sources with high quantum yield and photostability in quantum computing [7]. Furthermore, the combination of graphene and ND-s is in early stages of scientific interest. Said composites have been featured for uses in supercapacitor electrodes [8] and as a scanning probe in optical microscopy of graphene [9].

The mechanism of ND self-assembly onto graphene surface via electrostatic attraction is based on observations in several earlier works showing the possibility of attracting small (d~5 nm) detonation ND-s to graphene surface via appropriate pretreatment of the absorbates and graphene. It has been shown that nanoparticles tend to obtain the potential of the substrate [10] and also, there is a certain pH window for electrostatic attraction [11]. One of the goals of the work is to observe how well bigger ND-s (d~40 nm) function in similar conditions.

2. Experimental procedure

The ND-s used in this study were commercially available (Adamas Nanotechnologies Inc) detonation ND solution in water, containing 0.1% ND-s with average d~40 nm, zeta potential +30 mV, containing 1-4 N-V centers. The ND-s were first treated with H_2 plasma at 100 W power for 15 min (Diener Electronic Plasma Surface Technology Femto) to enhance positive zeta potential [5]. Then, the ND-s were dispersed in methanol via ultrasound (40 W, 30 min).

The substrate used was commercially available graphene on SiO_2/Si surface (Graphene Square Inc). To obtain negative zeta potential, the graphene surface was treated with O_2 plasma at 25 W for 15/30/60/120 s followed by dipping the treated substrate vertically into the ND-methanol solution for 1 minute.

It is expected that treatment with hydrogen and oxygen plasmas leads to vastly different zeta potentials, since oxygen plasma forms negatively charged surface (O^- terminated) whereas hydrogen plasma forms positively charged surface (H^+ terminated). By dispersing ND-s in methanol, alcyl groups are removed from the ND surface, leaving it hydroxylated, which leads to positive zeta potential [12]. Oxidized surface has negative potential, so the ND self-assembly is caused by electrostatical attraction. The efficiency of the surface covering and plasma treatment effect were estimated with scanning electron microscopy (SEM, type FEI Helios).

Additionally, luminescence (532 nm laser excitation, custom micro-luminescence setup) and Raman scattering measurements (Renishaw inVia Raman spectrometer) were carried out to estimate the applicability of optical detection of ND-s and graphene quality before and after plasma treatments.

3. Results and discussion

In Raman spectra of the plasma treated graphene (figure 1) the FWHM of 2D band (29 cm^{-1}) and intensity ratio of 2D/G (~2.8) show that the initial graphene is monolayer and the absence of D band hints good crystallinity. Relatively featureless graphene surface is also seen under optical microscope (figure 2). After treating the surface with O_2 plasma, high concentration of defects (or graphene edges) is indicated by an intense D band. Randomly distributed impurities and surface charges occurring during plasma treatment in graphene cause the G band to split into G and D' bands [13]. The merging of D' band into the G peak also implies high concentration of defects and the intensity ratio of D and D' band of ~3 indicates that the most common defects present are graphene edges [2].

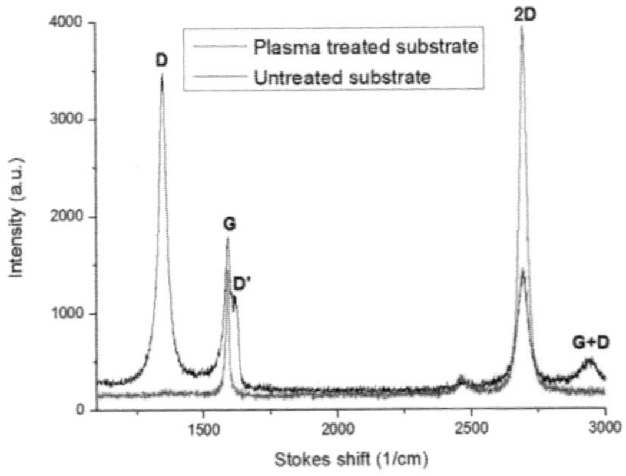

Figure 3. Raman spectra of untreated substrate and substrate treated in O_2 plasma for 15 s at 25 W.

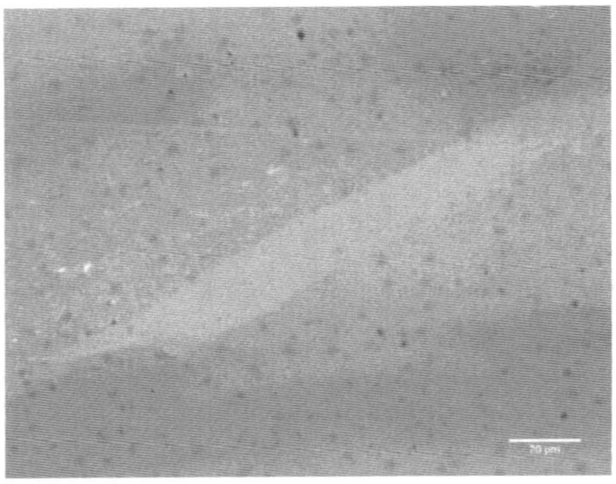

Figure 2: Optical image of untreated graphene.

D band observed after O_2 plasma treatment has been suggested to arise from two competing processes: (a) introduction of covalently bonded oxygen (epoxy, hydroxyl and carboxyl groups) onto the graphene lattice; (b) progressive amorphization of the sp^2 lattice [14]. Since the D'/D intensity ratio is small, the latter is less dominant in our case [2].

ND-s tend to accumulate at physical defects in graphene structure (figure 3). This is most likely due to the fact that the electrical field is concentrated onto sharp graphene features during surface plasma treatment and that defects are chemically more active, leaving them more/highly functionalized.

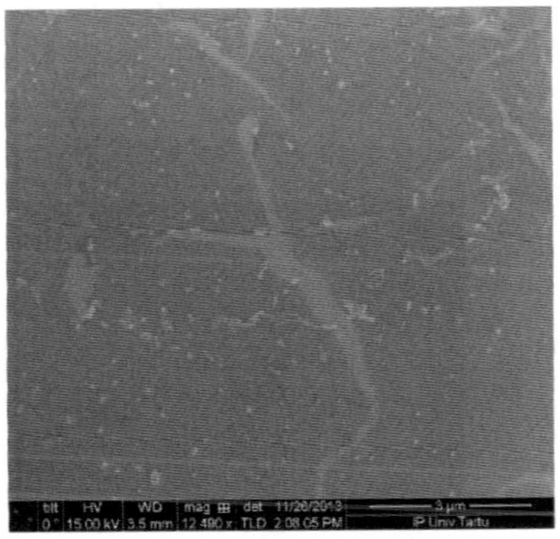

Figure 3. SEM of graphene-ND sample with 15 s 25 W O2 plasma surface treatment.

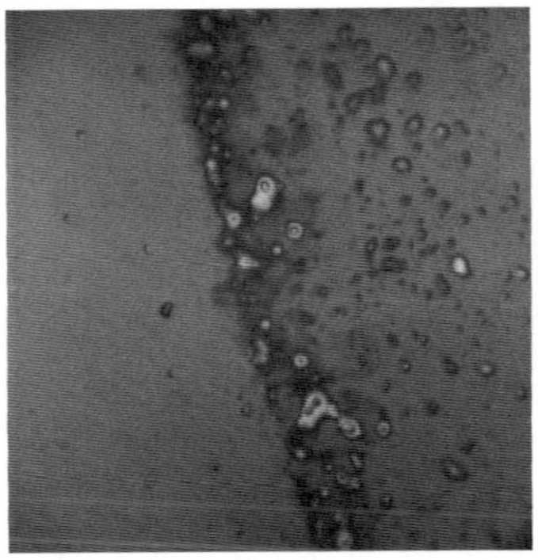

Figure 4. optical image of an edge of a droplet of ND-s.

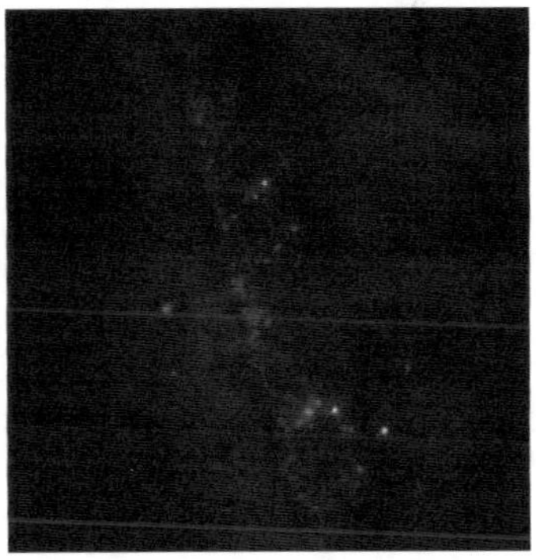

Figure 5. Fluorescence image of the same droplet (excitation λ=532 nm).

Single ND-s are not fluorescent on the graphene substrate due to fluorescence resonance energy transfer (FRET). Previous works have shown that single N-V center has lower fluorescence intensity and shorter lifetime on graphene compared to quartz due to FRET [15]. Fluorescence occurs only when there is a larger cluster of ND-s present on the graphene surface, for example at the edges of solvent droplet (figures 4 and 5).

Although the substrate shows significant loss of graphene layer as the plasma treatment time increases, ND-s still accumulate in the same places. ND-s tend to attach to graphene boundaries, regardless of progressive plasma etching (figure 6).

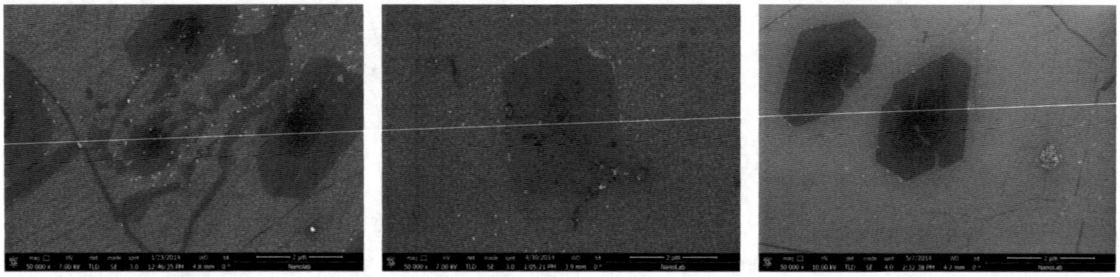

15 s 30 s 120 s

Figure 6. SEM micrograph of graphene-ND samples with different duration of O_2 surface plasma treatments. NDs attach preferably to graphene island edges (a defect).

The process of ND self-assembly by electrostatic attraction onto pretreated graphene surface is relatively random and challenging to control. Hees et al. [11] have shown that using centrifugation in addition to ultrasonic mixing and coating the substrate in an ultrasonic bath lead to an even covering of silicon with ND-s (RMS roughness 2 nm). In addition, their measurements show that hydrogenation leads to smaller degree of agglomeration, decreasing the mean particle size in solution from 100 nm to 3 nm. In our case the hydrogenation process had a smaller effect, SEM observations showed that there were no ND-s with diameter larger than 40 nm on the substrate, although they were present in the untreated solution.

The estimation of the optimum parameters for selective coverage of substrate's particular defects is highly complicated, most problematic being the determination of the potential between the intended target defect and the nanoparticle. Stehlik et al. [10] report that the reliable potential characteristic of a nanoparticle on any substrate is more dependent on the substrate than the particle's chemical termination and manifests as a linear function of the adhered particle's size. This means that already bound particles can alter the potential of the target during the electrostatic adhesion process, thus rendering simple zeta potential difference as a temporary indicator of attraction strength.

4. Conclusion

In this work we demonstrated a simple method for covering graphene surface defects with ND-s. The electrostatic attraction of different graphene defects and ND-s lead to accumulation of ND-s onto different micro-scale defects on graphene surface after plasma treatment. The coverage was only observable via SEM micrographs as the ND-s were not luminescent on graphene due to its quenching effect.

Current results suggest that zeta potential manipulation is a usable but complex tool for controlling ND immobilization on graphene. Further work using larger ND-s, preferably with dielectric shell, is needed to enable luminescent detection of graphene defects.

Acknowledgements

This work is supported by Estonian Nanotechnology Centre (project EU29996) and SA Archimedes project "Carbon Nanotube Reinforced Electrospun Nano-fibres and Yarns" (3.2.1101.12-0018) and by Estonian Nanotechnology Research Centre (NanoTAK). Authors would like to thank dr. I. Sildos for provided help in acquiring the ND samples and dr. V. Kiisk for assisting in spectroscopic measurements.

References

[1] Kamat P V. 2010 Graphene-Based Nanoarchitectures. Anchoring Semiconductor and Metal Nanoparticles on a Two-Dimensional Carbon Support *J. Phys. Chem. Lett.* **1** 520–7

[2] Eckmann A, Felten A, Mishchenko A, Britnell L, Krupke R, Novoselov K S and Casiraghi C 2012 Probing the nature of defects in graphene by Raman spectroscopy. *Nano Lett.* **12** 3925–30

[3] Denis P a. and Iribarne F 2013 Comparative Study of Defect Reactivity in Graphene *J. Phys. Chem. C* **117** 19048–55

[4] Geim a K and Novoselov K S 2007 The rise of graphene. *Nat. Mater.* **6** 183–91

[5] Ferrari A C, Meyer J C, Scardaci V, Casiraghi C, Lazzeri M, Mauri F, Piscanec S, Jiang D, Novoselov K S, Roth S and Geim A K 2006 Raman spectrum of graphene and graphene layers. *Phys. Rev. Lett.* **97** 187401

[6] Prabhakar N, Näreoja T, von Haartman E, Karaman D Ş, Jiang H, Koho S, Dolenko T a, Hänninen P E, Vlasov D I, Ralchenko V G, Hosomi S, Vlasov I I, Sahlgren C and Rosenholm J M 2013 Core-shell designs of photoluminescent nanodiamonds with porous silica coatings for bioimaging and drug delivery II: application. *Nanoscale* **5** 3713–22

[7] Greentree A D, Fairchild B A, Hossain F M and Prawer S 2008 Diamond integrated quantum photonics *Mater. Today* **11** 22–31

[8] Wang Q, Plylahan N, Shelke M V., Devarapalli R R, Li M, Subramanian P, Djenizian T, Boukherroub R and Szunerits S 2014 Nanodiamond particles/reduced graphene oxide composites as efficient supercapacitor electrodes *Carbon N. Y.* **68** 175–84

[9] Tisler J, Oeckinghaus T, Stöhr R J, Kolesov R, Reuter R, Reinhard F and Wrachtrup J 2013 Single defect center scanning near-field optical microscopy on graphene. *Nano Lett.* **13** 3152–6

[10] Stehlik S, Petit T, Girard H a, Arnault J-C, Kromka A and Rezek B 2013 Nanoparticles assume electrical potential according to substrate, size, and surface termination. *Langmuir* **29** 1634–41

[11] Hees J, Kriele A and Williams O a. 2011 Electrostatic self-assembly of diamond nanoparticles *Chem. Phys. Lett.* **509** 12–5

[12] Krueger A and Lang D 2012 Functionality is Key : Recent Progress in the Surface Modification of Nanodiamond *Adv. Funct. Mater.* 890–906

[13] Nourbakhsh A, Cantoro M, Vosch T, Pourtois G, Clemente F, van der Veen M H, Hofkens J, Heyns M M, De Gendt S and Sels B F 2010 Bandgap opening in oxygen plasma-treated graphene. *Nanotechnology* **21** 435203

[14] Nourbakhsh A, Cantoro M, Vosch T, Pourtois G, Clemente F, Veen M H Van Der, Hofkens J, Heyns M M, Gendt S De and Sels B F Bandgap opening in oxygen plasma-treated graphene **435203**

[15] Liu X, Wang G, Song X, Feng F, Zhu W, Lou L, Wang J, Wang H and Bao P 2012 Energy transfer from a single nitrogen-vacancy center in nanodiamond to a graphene monolayer *Appl. Phys. Lett.* **101** 233112

Visible photoluminescence of color centers in LiF crystals for absorbed dose evaluation in clinical dosimetry

J E Villarreal-Barajas[1,2,*], M Piccinini[2], M A Vincenti[2], F Bonfigli[2], R F Khan[1], R M Montereali[2]
[1] University of Calgary, 1331 29 St. NW, T2N 4N2, Calgary AB, Canada
[2] ENEA C.R. Frascati, Via E. Fermi 45, 00044 Frascati (Rome), Italy

*E-mail: eduardov@cancerboard.ab.ca

Abstract. Among insulating materials, lithium fluoride (LiF) has been successfully used as ionizing radiation dosemeter for more than 60 years. Thermoluminescence (TL) has been the most commonly used reading technique to evaluate the absorbed dose. Lately, optically stimulated luminescence (OSL) of visible emitting color centers (CCs) has also been explored in pure and doped LiF. This work focuses on the experimental behaviour of nominally pure LiF crystals dosemeters for 6 MV x rays at low doses based on photoluminescence (PL) of radiation induced CCs. Polished LiF crystals were irradiated using 6 MV x rays produced by a clinical linear accelerator. The doses (absorbed dose to water) covered the 1-100 Gy range. Optical absorption spectra show stable formation of primary F defects up to a maximum concentration of $2x10^{16}$ cm^{-3}, while no significant M absorption band at around 450 nm was detected. On the other hand, under Argon laser excitation at 458 nm, PL spectra of the irradiated LiF crystals clearly exhibited the characteristic F_2 and F_3^+ visible broad emission bands. Their sum intensity is linearly proportional to the absorbed dose in the investigated range. PL integrated intensity was also measured using a conventional fluorescence optical microscope under blue lamp illumination. The relationship between the absorbed dose and the integrated F_2 and F_3^+ PL intensities, represented by the net average pixel number in the optical fluorescence images, is also fairly linear. Even at the low point defect densities obtained at the investigated doses, these preliminary experimental results are encouraging for further investigation of CCs PL in LiF crystals for clinical dosimetry.

1. Introduction

Among insulating materials, lithium fluoride (LiF) in pure and doped form has been successfully used as clinical dosemeter based on thermoluminescence (TL) for over 60 years. In the last decade Al_2O_3:C, used as an optical stimulated luminescence (OSL) dosemeter, has been consolidated as a cost-effective clinical and personal dosimetry system [1]. OSL provides fast evaluation (ms), high sensitivity, wide dose range and dose image capabilities. These desirable dosimetric properties have encouraged research aimed to investigate the use of LiF as a photoluminescence (PL) based dosemeter [2]. Among aggregated color centers (CCs) produced by ionizing radiation in LiF, F_2 and F_3^+ defects, which consist of two electrons bound to two and three close anion vacancies, respectively, have almost overlapping absorption bands (M band) centered around 450 nm [3] and, therefore, can be simultaneously excited with a single pump wavelength in the blue spectral interval. On the other hand, they exhibit two different broad emission bands in the green (F_3^+) and red (F_2) spectral ranges [4]. In the last decades they were proposed for novel solid state luminescent detectors at high spatial resolution for soft x rays [5] based on LiF crystals and thin films [6]. Recently, their use was extended to proton beam characterization [7].

In this work the preliminary results of the optical investigation of radiation-induced CCs in 6 MV x ray irradiated pure LiF crystals in the clinically relevant dose range of 1-100 Gy are presented. Even at these low doses, optical absorption and PL spectra were measured at room temperature (RT) and a conventional optical microscopy based reading system has been successfully used for the integrated visible PL signal evaluation.

2. Materials and Methods

A set of nominally pure, commercially available, LiF crystals, of dimensions 5x5x0.5 mm^3, polished on both sides, were irradiated under full electronic equilibrium conditions using 6 MV x rays produced by a clinical linear accelerator at the Tom Baker Cancer Centre, Calgary, Canada. A 10x10 cm^2 field size was set for all the irradiations. The LiF crystals were positioned at the center of the square radiation field. The irradiations were set to 1, 10, 20, 50 and 100 Gy. All the doses refer to doses to water. After the irradiation, the LiF crystals were kept in the dark, but they were not protected from environment room light exposure during the optical absorption and PL measurements. The PL measurements were performed at RT and in a period of time after the irradiations that spanned from 6 weeks to 6 months.

The optical absorption measurements of the irradiated and blank (unirradiated) LiF crystals were performed using a Perkin-Elmer Lambda 950 spectrophotometer (Waltham, MA, USA) at normal incidence. The spectral range was set to 190–800 nm with a 1 nm resolution.

The laser induced PL spectra measurements were performed at RT using a continuous wave mode Argon laser at a fixed power of 25 mW. The laser was tuned at the wavelength of 458 nm in order to simultaneously excite the F_2 and F_3^+ CCs emission. Their photoemission was spectrally analyzed in the 480-800 nm range by a monochromator (Horiba Jobin Yvon, Triax 320) equipped with a grating blazed at 500 nm, with a 2 nm resolution and detected by a photomultiplier (Hamamatsu H7422-50) using a lock-in technique. A computer controlled acquisition system developed at ENEA Frascati was used for all the PL measurements. It is worth noting that all the PL spectra were corrected for the instrumental spectral response.

The integrated PL signal emitted by the irradiated LiF crystals was measured using a confocal optical microscope (Nikon Eclipse 80i-C1) working as a conventional wide-field optical microscope in fluorescence mode. The excitation light source was a mercury lamp OSRAM HBO 103W/2 (power 100W). The light of the mercury lamp (excitation) as well as the fluorescence (PL signal) of the LiF crystal were properly filtered. The excitation optical filter transmits light in the 450 – 490 nm spectral interval, while the output one (reading) allows the collection of light at wavelengths higher than 520 nm. The used two-dimensional detector for the PL signal acquisition was a sCMOS camera (Andor Neo), 2560x2160 pixels, pixel size 6.5 micron, front illuminated and cooled at -30°C. The PL integrated signal from all the irradiated crystals was obtained by directly extracting the average pixel number (gray level) of a circular area of approximately 0.75 mm^2. The freely available image processing software imageJ was used for all the image-PL analysis. All the images were converted to JPEG format and therefore the pixel numbers were limited to the 0-255 level, even if the used sCMOS resolution was 11 bit.

3. Results and Discussion

The absorption spectra, in optical density (O.D.), of a blank LiF crystal and of the LiF crystals irradiated at 10, 50 and 100 Gy by 6 MV x rays are shown in figure 1. Only these results are presented for the sake of clarity. The F absorption band, due to the primary F CCs, located around 250 nm, is clearly observed for the 50 and 100 Gy dose levels only. The M absorption band, due to F_2 and F_3^+ CCs, centred around 450 nm, is not detectable even for the highest doses used in this investigation [5]. The evaluation of the electronic defect concentrations inside the irradiated LiF crystals, was performed by using the Smakula formula [8]. For each spectrum, the O.D. of the blank LiF crystal was subtracted, in order to highlight the effects of coloration. The full width at half-maximum (FWHM) and the absorption coefficient at the band peak of the F absorption band were obtained by using a Gaussian best fit-procedure. The resulting concentrations of F CCs were 1.25×10^{16} and 1.95×10^{16} cm^{-3} for the LiF crystals irradiated at 50 and 100 Gy, respectively. For the M band, the absorption signal is comparable with the measurement noise. On the base of the Smakula formula, it was possible to obtain a rough estimation of the maximum concentration of F_2 CCs induced by the irradiation. In this case, the FWHM known from literature [8], and the absorption coefficient at 450 nm (around the band peak of the M absorption band) were used. The concentration of aggregate F_2 CCs was below 6×10^{14} cm^{-3}.

The PL spectra shown in figure 2 clearly present the two characteristic broad emission bands of the aggregate F_2 and F_3^+ defects. A background correction was applied: the net PL spectra were obtained by subtracting a blank spectrum (the spectrum of an unirradiated LiF crystal, used as "zero" dose reference) from every measured photoemission spectrum obtained for each irradiated LiF crystal. The F_2 and F_3^+ broad emission bands are centred around 678 and 541 nm, respectively. After converting the wavelengths from nm to eV, the rescaled PL spectra were fitted as the sum of two Gaussian bands (see figure 3), whose spectral parameters (peak position and FWHM) are in agreement with well-assessed literature in LiF crystals [9]. The areas under the Gaussian bands are associated with the contribution to the integrated PL of the F_2 and F_3^+ CCs. The integrated PL signal, defined as the sum of the PL signals from the two Gaussian bands, was obtained for every spectrum and plotted against the absorbed dose. These results are presented in figure 4. A power function (solid line in figure 4) was used for the fit. The exponent value of 0.99 confirms the linear behaviour of the PL response as a function of the absorbed dose.

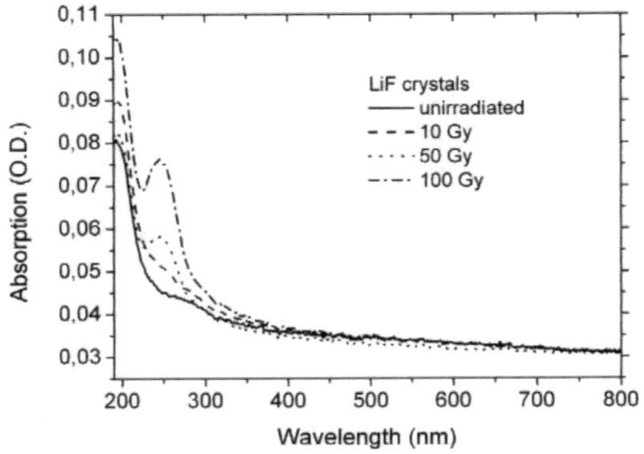

Figure 1. RT optical absorption spectra of 6 MV x rays irradiated LiF crystals at several doses. The spectrum of a blank crystal is reported for comparison.

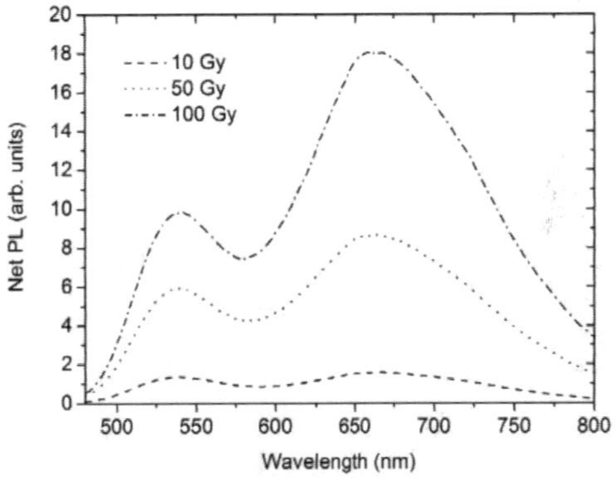

Figure 2. RT laser induced (458 nm) PL spectra of 6MV x-ray irradiated LiF crystals at 10, 50 and 100 Gy.

In the fluorescence microscope integrated PL signal measurements, in order to account for potential variations on the lamp intensity and/or sCMOS sensitivity, an unirradiated LiF crystal was always positioned side by side close to the irradiated LiF crystal, as illustrated in figure 5 for the 50 Gy x rays irradiated LiF crystal. Under this arrangement, the optical images of two identical circular areas set at a distance of 0.35 mm from the middle line (border line between the unirradiated and irradiated LiF crystal), were sequentially acquired in identical conditions. The difference between the mean pixel number of the irradiated crystal and the corresponding zero dose pixel number is defined as the net pixel number, which is associated with the PL signal of the irradiated crystal.

The integrated PL signal derived from the images of the irradiated crystals at all doses is presented in figure 6. Again it shows a linear behaviour in the investigated dose range (exponent 0.96), in agreement with the PL vs dose plot in figure 4, derived from the PL spectra. The error bars associated with the net pixel number estimates varied from 1.1 to 2.0 pixel number (standard deviation of the pixel number within the circular area). The percentage error is of the order of 4% at a dose of 100 Gy and rises at about 50% for 1 Gy irradiation and it is mainly due to the relatively high noise and low PL signals from the irradiated crystals. A significant increase in the sensitivity of the fluorescence microscope acquisition is required in order to effectively evaluate doses below 10 Gy level.

IOP Conf. Series: Materials Science and Engineering **80** (2015) 012020 doi:10.1088/1757-899X/80/1/012020

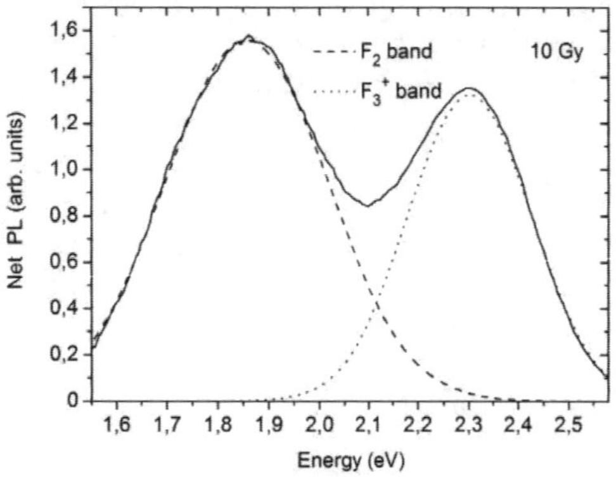

Figure 3. Net PL spectrum of the 6 MV x-ray irradiated LiF crystal at 10 Gy, best-fitted as the sum of two Gaussians bands, ascribed to the F_2 and F_3^+ CCs emission bands.

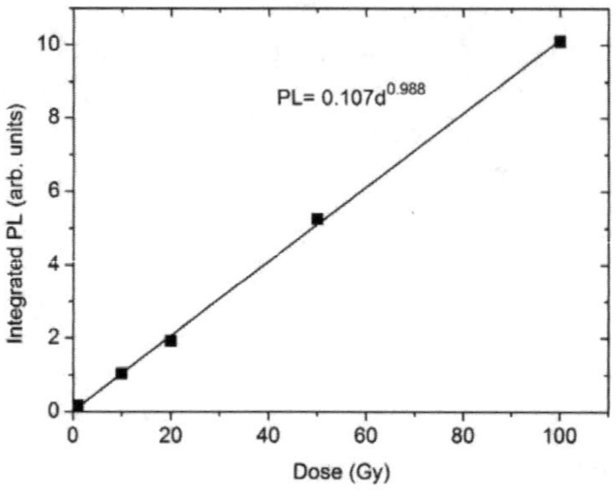

Figure 4. Integrated PL intensity as a function of the absorbed dose for all the 6 MV x rays LiF irradiated crystals and linear best fit (solid).

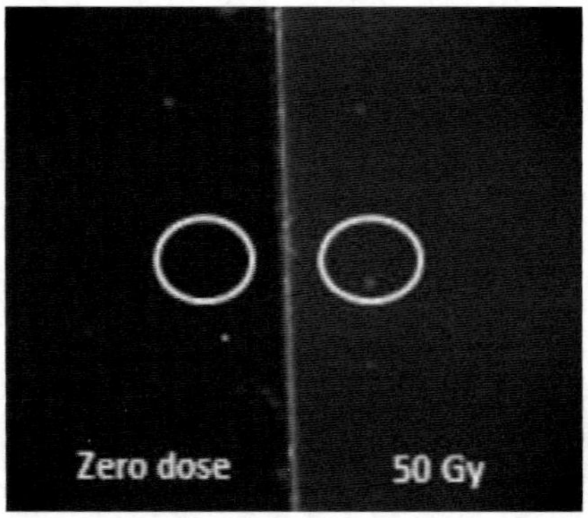

Figure 5. Fluorescence microscope image of a blank (zero dose) and the 50 Gy 6 MV x ray irradiated LiF crystals. The two white circles show the areas selected for the measurement of the integrated PL signal.

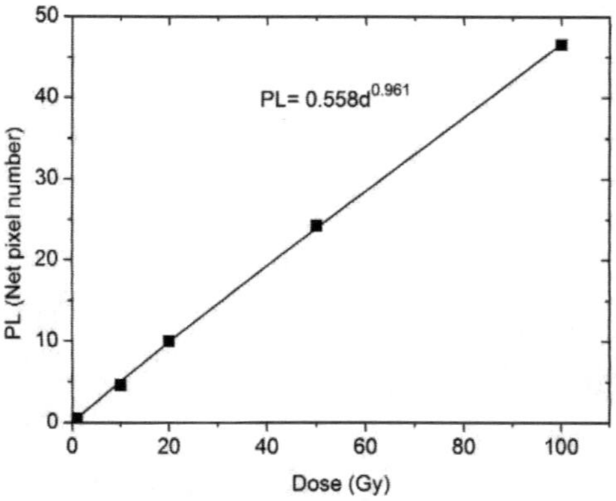

Figure 6. Fluorescence microscope integrated PL signal (Net pixel number) as a function of the absorbed dose for all the 6 MV x-ray irradiated LiF crystals.

4. Conclusions and further work

The use of pure LiF crystals as PL dosemeters in the clinically relevant 1 to 100 Gy dose range has been investigated for 6 MV x rays irradiation. The linearity of the integrated PL response of F_2 and F_3^+ electronic defects as a function of dose, derived from the laser induced visible PL spectra, is consistent with the optical response detected with a fluorescence microscope system. This linearity is a very desirable feature of any radiation detector. Even at the low electronic defect densities (F and F_2 concentrations below $2x10^{16}$ cm^{-3} and $6x10^{14}$ cm^{-3}, respectively) obtained at the investigated doses, these preliminary experimental results grant further investigation of CCs PL in LiF crystals for clinical dosimetry. A modified excitation light system as well as a more effective optical filtering of the excitation and emitted PL from the LiF irradiated crystals is under optimization, in order to address the low sensitivity of the fluorescence microscope reading system. Re-usability of the LiF crystals and laser induced PL reproducibility after thermal annealing requires additional investigation.

References

[1] Yukihara E G and McKeever S W S 2008 *Phys. Med. Biol.* **53** R351
[2] Bilski P, Marczewska B, Twardak A, Mandowska E and Mandowski A 2014 *Radiat. Meas.* (in press)
Matsushima L C, Veneziani R G and Campos L L 2013 *Radiat. Meas.* **56** 365
Oster L, Druzhyna S, Orion I and Horowitz Y S 2013 *Radiat. Meas.* **56** 320
Marczewska B, Bilski P, Mandowska E and Mandowski A 2012 *Cent. Eur. J. Phys.* **10** (4) 1009
Oster L, Druzhyna S and Horowitz Y S 2011 *Nucl. Instr. Meth.* A **648** 261
[3] Nahum J and Wiegand D A 1967 *Phys. Rev.* **154** 817
[4] Vincenti M A, Marolo T and Montereali R M 2010 *Radiat. Meas.* **45** 359
[5] Baldacchini G *et al.* 2005 *Rev. Sci. Instrum.* **76** 113104_1
[6] Almaviva S, Bonfigli F, Franzini I, Lai A, Montereali R M, Pelliccia D, Cedola A and Lagomarsino S 2006 *Appl. Phys. Lett.* **89** 54102
[7] Piccinini M, Ambrosini F, Ampollini A, Carpanese M, Picardi L, Ronsivalle C, Bonfigli F, Libera S, Vincenti M A and Montereali R M 2014 *Nucl. Instr. Meth.* B **326** 72
[8] Vincenti M A, Baldacchini G, Kalinov V S, Montereali R M and Voitovich A P 2010 *IOP Conference Series: Materials Science and Engineering* **15** 012053
Fowler W B 1968 *Physics of Color Centers* (New York and London: Academic Press)
[9] Baldacchini G, De Nicola E, Montereali R M, Scacco A and Kalinov V 2000 *J. Phys. Chem. Solids* **61** 21

IOP Conf. Series: Materials Science and Engineering **80** (2015) 012021 doi:10.1088/1757-899X/80/1/012021

Luminescence properties of the Pb_2MoO_5 single crystals doped with Cu ions

V. Chornii, S. Nedilko

Taras Shevchenko National University of Kyiv, 01601, Kyiv, Ukraine

E-mail: vchornii@gmail.com

Abstract. Photoluminescence (PL) emission and PL excitation spectra of the undoped and Cu-doped Pb_2MoO_5 grown by Czochralski technique are obtained and analyzed. The complex wide PL emission band in 450-750 nm spectral range was observed for both pure and Cu-containing crystals under photoexcitation with ultraviolet light. The short-wavelength component of the PL has been associated with luminescence centers related to MoO_4^{2-} anionic groups. Long-wavelength the PL component with maxima near 600 nm was observed for both pure and Cu-doped crystals and it is described to manifestation of the crystal defects.

1. Introduction

Optical properties of the double lead molybdate Pb_2MoO_5, single crystals have been studied due to the perspectives of this material application in the acousto-optic devices [1, 2]. In the same time due to the photochromic effect, peculiar to the lead molybdate family, their application is still limited [3]. As it follows from the literature both Mo [4] and Pb deficiency [5] can lead to a photochromic effect in $PbMoO_4$, for example. Obviously, that non-stoichiometric composition of the crystals promotes the creation of the various types of defects in the crystals structure. The situation can be partially managed by the specific doping of the crystals with impurity ions. The dopants will affect the quantity of the defects and influence their role. As was reported, the $PbMoO_4$ doping with Bi^{3+} and Ba^{2+} ions leads to lowering of the photo-induced coloring of the compound [3]. So, it is possible that isovalent to Pb^{2+} ion doping can change an optical quality of the Pb_2MoO_5 crystals. Therefore the role of the defects in the optical properties determination is an important issue for developing of high quality optical materials based on the Pb_2MoO_5 crystals.

Optical methods, in particular luminescence spectroscopy, can be efficiently used to analyze the role of defects on the physical processes in oxide crystals. Luminescence properties of the undoped and impurity-containing molybdates are interesting in itself for elaboration of the scintillation and luminescent materials. To our knowledge, there are only two papers on luminescence properties of the nominally un-doped Pb_2MoO_5 crystals [6, 7], while no paper deals with luminescence of the impurity-containing double lead molybdate.

In the present paper we studied the luminescence properties of both pure and Cu-doped Pb_2MoO_5 crystals and have tried to find an influence caused by Cu ions. The possible origin of the investigated crystals luminescence is also discussed.

2. Sample preparation and experimental set-up

The Pb_2MoO_5 single crystals were grown from melts in a Pt crucible by the conventional Czochralski technique. The melts were prepared by heating of PbO and MoO_3 oxides (purity rating 99.99%) taken in stoichiometric ratio. Typical growth conditions are the following: 50 rpm rotation rate, 3 mm/h pulling rate, 30°C/h cooling rate. The samples were grown in ambient air atmosphere with temperature fluctuation during crystal growing within ±0.2°C. The Pb_2MoO_5:Cu crystals were synthesized in the same way from a melt that contains 0.03 wt. % of the CuO.

The PL emission and PL excitation spectra in the ultraviolet (UV) region of excitations were measured using a Xenon lamp DKsSh-150, primary monochromator MDR-12 and secondary monochromator MDR-23. The samples were put into a helium-flow cryostat for the luminescence study. The PL emission and excitation spectra were corrected on system response.

3. Results and discussion

The PL emission spectra of Pb_2MoO_5 and Pb_2MoO_5:Cu crystals at liquid helium temperature (4.2 K) are presented on the Figure 1. It is clearly seen from the figure that both pure and Cu-doped crystals reveal a wide complex PL band in the 450 - 750 spectral region. At least, two components can be distinguished in the PL spectra of the un-doped Pb_2MoO_5 crystals under long-wave ultra-violet (UV) excitation at λ_{ex} = 350 nm.

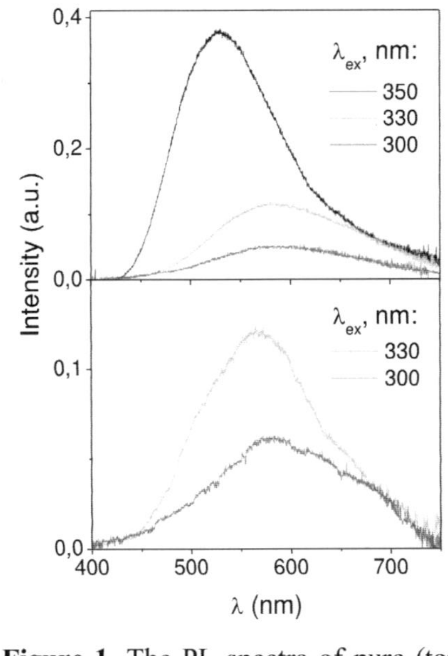

Figure 1. The PL spectra of pure (top) and Cu-doped Pb_2MoO_5 crystals (bottom); T = 4.2 K.

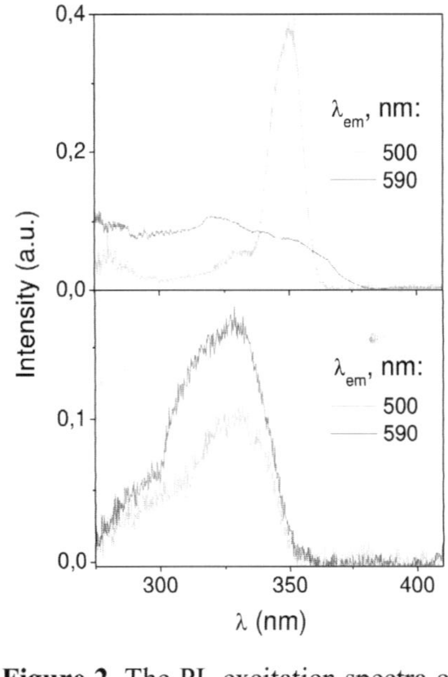

Figure 2. The PL excitation spectra of pure (top) and Cu-doped Pb_2MoO_5 crystals (bottom); T = 4.2 K.

The short-wavelength ("green") PL component has a maximum at 520 nm and was observed earlier [6]. This green PL is mainly excited in the narrow excitation band with maxima near 350 nm (3.54 eV) in excitation spectra. This asymmetric PL excitation band is accompanied with a shoulder in the range 275 - 340 nm and low intensity excitation band which is manifested as the wing located in the range 250 - 275 nm (Figure 2, top).

Figure 1 shows also the long-wavelength ("orange") component of the un-doped Pb_2MoO_5 photoluminescence with maximum near 590 nm. This PL clearly observed under excitation at shorter wavelengths, λ_{ex} = 330 and 300 nm, and the low intensity "red" component of the PL can be also distinguished in the range 625 - 750 nm. The mentioned orange and red components are mainly excited in the shorter UV wavelength range in comparison with the PL excitation of the green luminescence, 250 – 340 nm, but at the same time we have to note the excitation at the longer wavelengths range: 360 – 380 nm (see the curves on the Figure 2, top). We should note that only orange photoluminescence of the un-doped Pb_2MoO_5 crystal was earlier observed for excitation with synchrotron radiation in vacuum ultraviolet (VUV) spectral region ($\lambda_{ex} < 200$ nm) [7].

The knowledge of the band gap value, E_g, could simplify the analysis of the PL excitation spectra and the PL origin. Unfortunately no papers which deal with E_g of the Pb_2MoO_5 compound are available at present. However, we know that thin samples of the un-doped double lead molybdate crystals are transparent for spectral region $\lambda > 336$ nm and the region of the high absorption starts at 230 nm [8]. Taking into account the high absorption of most oxide materials in the region of the band-to-band transitions we have full reason to assume that the band gap of Pb_2MoO_5 is not lower that 3,7 eV (this value correspond to wavelength 336 nm). Because the maximum of the green PL excitation band (3.54 eV) is close to the above mentioned value, it probably indicates the "excitonic" origin of the green luminescence. It should be noted the green luminescence band position as well as it's excitation spectrum is similar to the case of the $PbMoO_4$ luminescence and it is generally accepted that luminescence of the $PbMoO_4$ is related to radiative decay of excitons which localized at molybdate groups of the crystal. So, it is likely that green component of the Pb_2MoO_5 PL has the same origin. This is a radiative decay of excitons which are localized at the MoO_4^{2-} molybdate groups.

As for the orange and red the PL components it is not so easy to speak about their origin. In fact, the VUV and UV excitations, obviously, correspond to band-to-band transition in Pb_2MoO_5 crystal and it would be useful to engage available data on the electronic band structure calculations. It has been found earlier, the distribution of the partial density of states of the Pb_2MoO_5 indicates that Pb-O chains contribute to the band-to-band absorption edge [7]. Therefore, mentioned chains or rather Pb^{2+} ions bounded with oxygen in its surrounding may play an important role in luminescence excitation. At the same time, the orange and red PL bands were observed for a number of molybdate and analogues tungstate compounds under excitation not only in VUV and UV regions, but in the region of crystals transparency and that is characteristics detail – in the spectral range located near excitonic excitation band at longer wavelengths [9-14]. That is why, those types of the PL are assigned to the various complex luminescence centers which can include two charged matrix cation (or its vacancy), molecular group (molybdate or tungstate), oxygen vacancy (or F, F^+, F^{++}, etc, – center formed on the site of the oxygen vacancy) [see e.g. 15]. Our measured excitation spectra (Figure 2, top) show the similar characteristic details, so we can assume that the orange and red components of the Pb_2MoO_5 PL can be assigned to the similar luminescence centers, but no more reliable conclusions about the origin of the orange and red luminescence of un-doped Pb_2MoO_5 crystals can be done on the basis of the data available at present.

Doping of the Pb_2MoO_5 crystal with Cu leads to the disappearance of the green photoluminescence band (Figure 1, b) and corresponding to it a band in the PL excitation spectra (Figure 2, b). The orange PL bands are still observed but have somewhat lower intensity if compared to the un-doped samples. As for the red PL, its contribution to the spectra can be distinguished clearly.

There are at least two possible ways of the Cu ions location in the Pb_2MoO_5 crystal matrix can lead to vanishing the green component of the PL by copper impurity. We have to regard two charge variant of the Cu substitution for the Pb^{2+} in crystal lattice. First of them is Cu^{2+} substitution for the Pb^{2+}. If the Cu^{2+} substitutes for the Pb^{2+} in crystal lattice the changes in nearest oxygen surrounding seem to be indispensable (the ionic radii of the Pb^{2+} and Cu^{2+} are strongly different: 1.26 and 0.80 Å respectively).

The other way is more complicated. This is related to other charge state of copper ion – Cu^+. It was reported for similar $PbWO_4$ compound that impurities of copper are mainly in the Cu^+ state, but not in the Cu^{2+} state [16]. We can expect a similar situation for the Pb_2MoO_5:Cu case. An incorporation of the Cu^+ ion on the Pb^{2+} site requires charge compensation, so, the additive oxygen vacancies have to be formed in the $Pb_2MoO_5:Cu^+$ crystals. Note, that ionic radii of the Pb^{2+} and Cu^+ are also different: 1.26 and 0.98 Å respectively. Possibility of the Cu^+ ions substitution for Pb^{2+} is in accordance with the results of the Pb_2MoO_5:Cu crystal electronic band structure calculations, which were performed by us recently with Wien2k package [17]. Of course, some doubts arise due to approximate character of calculation based on the density functional theory quantitative analysis of the state occupations. Nevertheless it is likely that in the Pb_2MoO_5 crystal the Cu^+ ions on Pb^{2+} positions can exist.

In both cases, some of the MoO_4^{2-} anionic groups in Pb_2MoO_5:Cu crystals are more distorted than in un-doped Pb_2MoO_5 that results in their losing the capability to localize excitation energy and future emit the "green" light as result of the excitons decay.

Orange and red components of the Pb_2MoO_5:Cu crystal luminescence are of the same origin as ones for the un-doped Pb_2MoO_5. Some changes in the shapes of the emission and excitation spectra of the orange and red luminescence can be related with changes of the various types of defects content and related with them centers of mentioned luminescence. This issue, charge state of the Cu ions as well as nature of the orange and red photoluminescence requires further research, in particular by time resolved spectroscopy, of both un-doped and doped with different content of the Cu ions Pb_2MoO_5 crystals.

4. Conclusions

Studied samples of the un-doped Pb_2MoO_5 crystals grown by Czochralski technique reveal complex photoluminescence bands with two main components maxima at 520 and 590 nm respectively. The PL excitation spectra of short-wavelength, "green", component reveal narrow band at 350 nm accompanied with a shoulder on the side of the lower wavelengths.

Doping of the Pb_2MoO_5 crystal with a low amount of Cu leads to vanishing of green photoluminescence as well as to disappearance of the band with maxima at 350 nm in the PL excitation spectra. This green luminescence was attributed to radiative decay of localized excitons.

The changes in luminescence properties of Cu-doped samples in comparison with pure crystals are ascribed to the increasing of the defects concentration (i.e. oxygen vacancies) in crystal lattice.

The origin of the orange and red long-wavelength luminescence of the Pb_2MoO_5 crystals requires further research.

Acknowledgments

The authors are grateful to M. P. Trubitsyn, the leader of the research team, the Head of Department of Solid State Physics and Optoelectronics of the Oles Honchar Dnipropetrovsk National University (Dnipropetrovsk, Ukraine) for the samples given to us for experiments.

References

[1] Ohmachi Y and Uchida N 1971 *J. Appl. Phys.* **42** 521-524
[2] Mil'kov M G, Volnyanskii M D, Antonenko A M and Voloshinov V B 2012 *Acoustical Physics* **58** 172-179
[3] Bochkova T M, Volnyansii M D, Volnyansii D M and Shchetinkin V S 2003 *Physics of the Solid State* **45** 244-247
[4] Neiman A Ya, Afanasiev A A, Feodorova L M, Gabrielian V T and Karagezian S M 1984 *Physica Status Solidi* A **83** 153-158
[5] van Loo W 1975 *J. Solid State Chem.* **14** 359-365
[6] Folkerts H F, Zuidema J and Blasse G 1996 *Solid State Commun.* **99** 655-658
[7] Nedilko S, Chornii V, Hizhnyi Yu, Trubitsyn M and Volnyanskaya I 2014 *Opt. Mater.* **36** 1754-1759
[8] Uchida N, Miyazawa S and Ninomiya K 1970 *J. Opt. Soc. Am.* **60** 1375-1377
[9] van Loo W and Wolterink D J 1974 *Phys. Lett.* **47A** 83-84
[10] Mazurak Zb, Blasse G and Liebertz J 1987 *J. Solid State Chem.* **68** 181-184
[11] Fujita M, Itoh M, Mitani H, Sangeeta and Tyagi M 2010 *Physica Status Solidi* **247** 405-410
[12] Itoh M 2012 *J. Lumin.* **132** 645-651
[13] Hizhnyi Yu, Nedilko S, Chornii V, Nikolaenko T, Zatovsky I, Terebilenko K and Boiko R 2013 *Solid State Phenomena* **200** 114-122
[14] Chukova O and Nedilko S 2013 *Opt. Mater.* **35** 1735-1740

[15] Annenkov A, Auffray E, Korzhik M, Leqoc P and Peigneux J-P 1998 *Physica Status Solidi* A **170** 47-62

[16] Laguta V V, Martini M, Vedda A, Rosetta E, Nikl M, Mihokova E, Rosa J and Usuki Y 2003 *Phys. Rev.* B **67** 205102-1-8

[17] Blaha P, Schwarz K, Madsen G, Kvasnicka D and Luitz J, 2001. *WIEN2k, An Augmented Plane Wave + Local Orbitals Program for Calculating Crystal Properties* (Wien, Austria: Karlheinz Schwarz, Techn. Universität) ISBN 3-9501031-1-2

IOP Conf. Series: Materials Science and Engineering **80** (2015) 012022 doi:10.1088/1757-899X/80/1/012022

Thermoluminescence in LiF crystals and the role of impurities

M A Vincenti[1], M Ambrico[2], G Baldacchini[3], V S Kalinov[4], R M Montereali[1], E Nichelatti[5], L Schiavulli[6], A P Voitovich[4]

[1]ENEA C.R. Frascati, UTAPRAD-MNF, Photonics Micro- and Nano-structures Laboratory, Via E. Fermi 45, 00044 Frascati, Rome, Italy
[2]CNR-IMIP UoS di Bari, Via Orabona, 4, I-70126 Bari, Italy
[3]ENEA guest, Via G. Quattrucci 246, 00046 Grottaferrata, Rome, Italy
[4]Institute of Physics, National Academy of Science of Belarus, Nezavisimosti Prospect 68, 220072 Minsk, Belarus
[5]ENEA C.R. Casaccia, UTTMAT-OTT, Optical Devices Laboratory,Via Anguillarese, 301, 00123 Santa Maria di Galeria, Rome, Italy
[6]Physics Department of University and Sezione INFN of Bari, Via Orabona, 4, I-70126, Bari, Italy

E-mail: maria_aurora.vincenti@enea.it

Abstract. Eight nominally pure lithium fluoride crystals, obtained from a single LiF crystal containing less than 100 ppm impurities, were irradiated by gamma-rays from a ^{60}Co source at room temperature, with doses from 8.4 to 2.5 x 10^5 Gy, but one at -60 °C. Optical density measurements were performed to investigate the radiation-induced color centres (CCs) and to evaluate their concentrations. Thermoluminescence (TL) glow curves were collected and simulated by a first-order kinetics approach, and from the best-fit procedure ten glow peaks (GPs) spanning from 100 to 450 °C were highlighted. A comparative analysis of GP intensities and CC concentrations as a function of the irradiation dose has questioned their association as obtained in previous measurements, showing the impurities, less than 10^{18} cm^{-3}, still playing a predominant role in TL spectra. New measurements on LiF crystals more pure, at least an order of magnitude, are required to establish for sure the association of GPs to CCs.

1. Introduction

Thermally stimulated light emission from an insulator or a semiconductor crystal, following the previous absorption of energy from ionizing radiation, results in the typical thermoluminescence (TL) glow curves, from which information on the electron traps present in the crystal band gap can be extracted [1]. TL is one of the most frequently used techniques in radiation dosimetry for personal and environmental use, with outstanding success [1, 2]. Among the various crystalline materials utilized as TL dosimeters, lithium fluoride (LiF) is the most common because of its almost biological tissue equivalence and its chemical-physical properties. LiF crystals are relatively hard and almost not hygroscopic. Irradiation of LiF crystals by ionizing radiation (X-rays, gamma-rays, protons, electrons, etc.) induces the stable formation of primary and aggregate color centers (CCs), whose optical and spectroscopic properties are rather well known [3]. The primary electronic defect is the F center, which consists of an anionic vacancy occupied by an electron. Its absorption band, called F band, is located at about 248 nm.Up to now, photoluminescence originating from the F center in LiF has not been detected unambiguosly. The F_2 and F_3^+ aggregate centers (two electrons bound to two and three anion vacancies, respectively) possess almost overlapping absorption bands at about 450 nm, which together form what is generally called M band [3]; under optical pumping in this spectral region they emit broad photoluminescence bands peaked at 678 nm and 541 nm for F_2 and F_3^+ centers, respectively [4]. The F_3 aggregate centers (three F centers, located at the vertices of an equilateral triangle in the lattice plane {1,1,1}) give rise to two principal absorption bands, $F_3(R_1)$ and $F_3(R_2)$, located at around 3.92 eV (316 nm) and 3.31 eV (374 nm), respectively [5].

Even though the TL properties of LiF have been extensively investigated since long times, the role of impurities and radiation-induced CCs in TL spectra [6] has still to be clarified. Indeed, the mechanism of TL is not known in detail yet, although a general description has been attempted with some success. This situation has been addressed lately having in mind the role of impurities versus the properties of pure LiF material [7, 8].

During the last decade systematic studies by TL technique have been carried out to establish the role of radiation-induced CCs in TL spectra, with promising results [6, 9]. In particular, four LiF crystals, irradiated with the same dose and properly treated in order to get a wide variation of CCs content, have been studied with interesting results [7]. In order to extend that latest investigation and clarify the role of CCs and impurities in TL glow curves, we started a new campaign of TL measurements on a larger number of LiF crystals and with a much wider range of irradiation doses. In this work some preliminary results are reported and discussed.

2. Materials and Methods

Eight nominally pure LiF samples, obtained from a single crystal, containing less than $10^{18}/cm^3$ impurity ions in total, were irradiated at increasing doses by gamma-rays from a ^{60}Co source. Seven samples were irradiated at room temperature (RT) with doses from 8.4 and $2.5x10^5$ Gy, and one with a dose of $8.4x10^4$ Gy while keeping its temperature at -60 °C (213 K).

Optical density measurements were performed at normal incidence and at RT, by using a double-monochromator Perkin-Elmer Lambda 950 spectrophotometer, in the 190-1400 nm spectral range with 1 nm of resolution.

TL was measured in the range 50-450 °C (323-721 K) with a heating rate of 0.25 °C/s (15 °C/min) by using a commercial automated RISØ TL/OSL-DA-15 reader, also equipped with a $^{90}Sr/^{90}Y$ beta (β) radiation source (0.565 MeV mean β energy, dose rate 0.119 Gy/s). The glow curves were measured from small fractions of powder obtained by pounding small pieces of the LiF crystals under a nitrogen flux. Five pieces of each irradiated LiF crystal, of about equal weight, were pounded and measured for statistical purposes, and the final TL glow curves were averaged by taking in account the weight of each powder sampling.

3. Results and Discussion

Figure 1 shows the optical density spectrum and its best fit of sample S15, which was irradiated at RT at a dose of $2.5x10^3$ Gy. The most visible spectral features are the F band, peaked at 5 eV (248 nm), and the M band located at about 2.76 eV (450 nm). The absorption band contributions of more complex CCs, such as $F_3(R_1)$, $F_3(R_2)$, $F_4(N_1)$ and $F_4(N_2)$ located at about 3.92 eV (316 nm), 3.31 eV (374 nm), 2.40 eV (517 nm) and 2.26 eV (547 nm), are also highlighted, although not observed in this sample.

Figure 2 shows an example of a TL glow curve (sample S15) and its best fit achieved by using the GlowFit free-downloadable software, which was especially designed to simulate TL glow curves resulting from up to ten single or superimposed glow peaks (GPs) following first order kinetics [10]. Comparing the results of the present investigation [11] with previous ones [6, 9], it appears that the GP temperatures differ in the two measurements by about 19 °C in average. This is due in part to unavoidable instrumental systematic errors and mostly to an improved best-fit procedure in the present case.

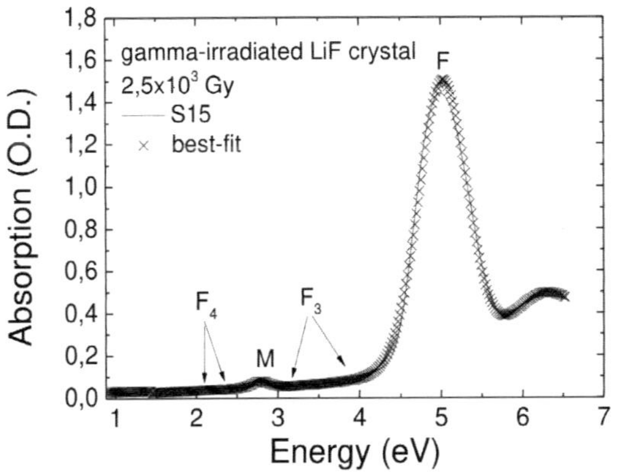

Figure 1. RT optical density spectrum of the gamma colored LiF crystal at a dose of 2.5x10^3 Gy, S15 sample, (black full line) and its best fit (x symbol). The thickness of the crystal is 2 mm.

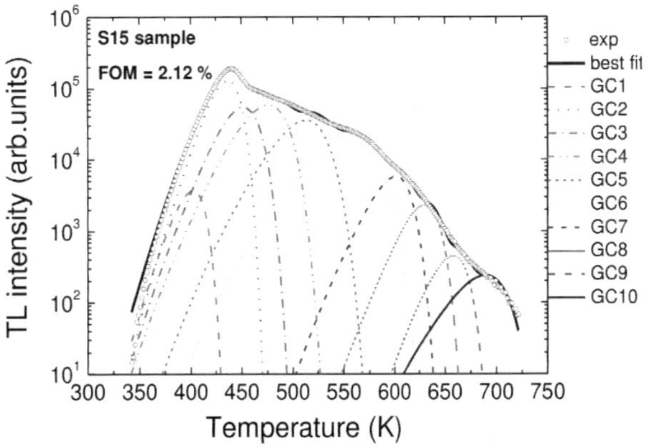

Figure 2. TL spectrum of the gamma colored LiF crystal at a dose of 2.5x10^3 Gy, S15 sample, (circle), its best fit (black full line) the ten glow peaks (colored lines). The Figure Of Merit (FOM) is indicated.

From optical density measurements it is possible to approximately calculate the amount of CCs in the irradiated LiF crystals [12]. Figures 3 and 4 report in a log-log scale the concentrations of F centers, and $F_3(R_1)$ and $F_3(R_2)$ centers, respectively, as functions of the irradiation dose. The experimental points have been interpolated with a line, except for sample S18 which was treated differently (dose of 8.4x10^4 Gy at -60 °C).

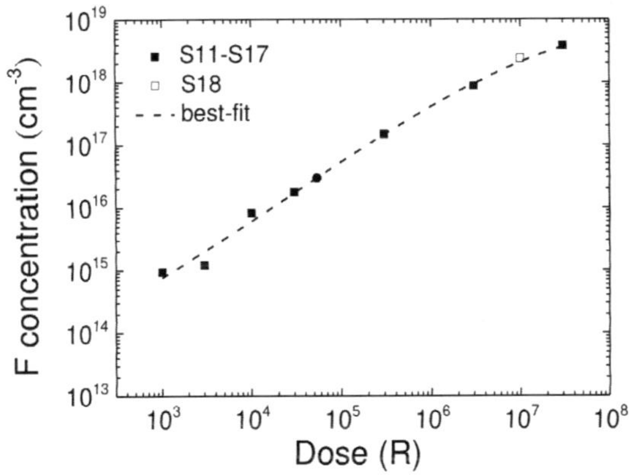

Figure 3. Concentration of F centers as a function of gamma ray dose for the LiF crystals colored at RT and utilized in the present study (1 R ≅ 8.3 10^{-3} Gy).

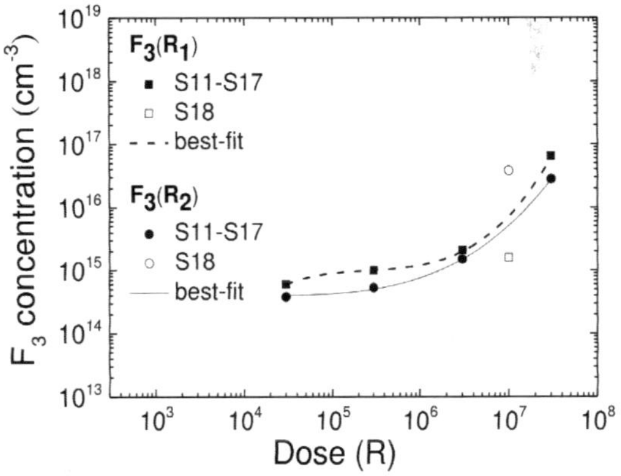

Figure 4. Concentration of $F_3(R_1)$ and $F_3(R_2)$ centers as functions of gamma ray dose for the eighth samples utilized in the present study. The three samples irradiated at the lowest doses do not give appreciable numerical values (1 R ≅ 8.3 10^{-3} Gy).

Figures 5 and 6 report the intensities of GP7, and GP3 and GP4, respectively, as functions of the irradiation dose. GP7 intensity shows a linear behavior with dose, thus it can be related to F centers, whose concentration is also linear with the dose, see figure 3, and moreover the sample S18 result is compatible with linear behavior in both graphs. In the previous analysis it was associated with F centers bound to impurities contained in LiF crystals [9], which cannot be excluded by current investigation.

In the previous analysis [9], GP3 and GP4, figure 6, were associated with F_3 centers, but examining the behavior of the F_3 concentrations as functions of dose, figure 4, this assignment cannot be confirmed. Indeed, the peak intensities of GP3 and GP4 grow together with the concentrations of $F_3(R_1)$ and $F_3(R_2)$, albeit with a different behavior, but, while the concentrations continue to grow, GP3 and GP4 intensities saturate. Moreover, also the values of the F_3 concentrations and the GP3 and GP4 intensities of sample S18 do not support this association. The observed saturation can be explained with the hypothesis that these GPs are due to the presence of CCs bound to impurities contained in the LiF crystals. Indeed, at a certain dose, there are not "enough impurities" to be bounded to CCs, and the related GP intensities cannot grow anymore. Similar behavior has been found in the other samples [11].

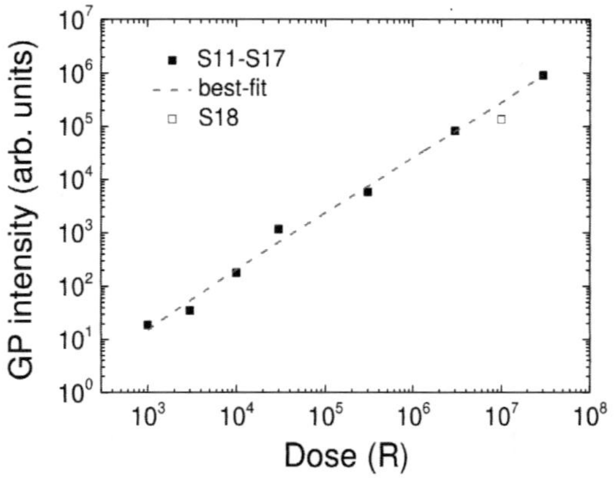

Figure 5. Intensity of GP7 as a function of gamma ray dose for the eighth samples utilized in the present study (1 R ≅ 8.3 10^{-3} Gy).

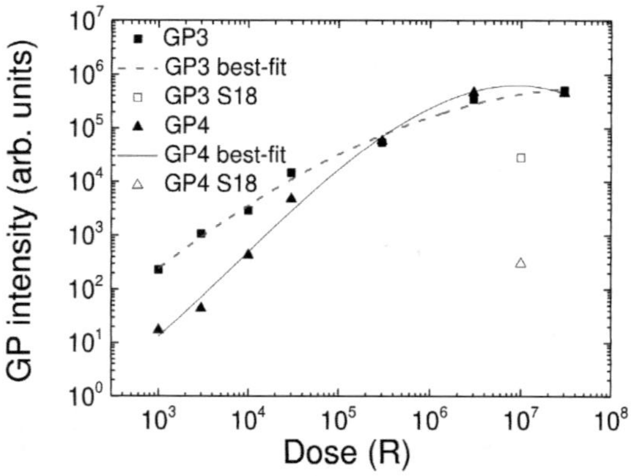

Figure 6. Intensity of GP3 and GP4 as a function of gamma ray dose for the eighth samples utilized in the present study (1 R ≅ 8.3 10^{-3} Gy).

4. Conclusions

The association between GPs and CCs performed a few years ago [4, 7] was based on the knowledge of the physics of CCs and accurate annealing measurements on a restricted set (only four samples) of LiF crystals that had been gamma-irradiated at the same dose, and so not completely satisfactory.

This new campaign of measurements, performed on a larger number of LiF crystals and with a much wider range of irradiation doses, has shown, in the framework of the first-order kinetics theory, that the previous association between CCs and GPs is strongly questionable and that impurities could be responsible for certain GPs that had been earlier assigned to CCs. These results highlight the paramount influence of residual impurities in LiF crystals, even if apparently negligible, in TL measurements. New measurements on LiF crystals, containing much less impurities, are required to establish on more sound bases the association of GPs to CCs.

References

[1] Chen R and Mc Keever S W S 1997 *Theory of Thermoluminescence and Related Phenomena*, World Scientific, Singapore

[2] Furetta C and Weng P S 1998 *Operational Thermoluminescence Dosimetry*, World Scientific, Singapore

[3] Nahum J and Wiegand D A 1967 *Phys. Rev.* **154** 817

[4] Baldacchini G, De Nicola E, Montereali R M, Scacco A, Kalinov V 2000 *J. Phys. Chem. Solids* **61** 21

[5] Montecchi M, Nichelatti E, Mancini A, Montereali R M 1999 *J. Appl. Phys.* **86** 3745

[6] Baldacchini G, Montereali R M, Nichelatti E, Kalinov V S, Voitovich A P, Davidson A T, and Kozakiewicz A G 2008 *J. Appl. Phys.* **104** 063712-1-10

[7] Davidson A T, Kozakiewicz A G, Wilkinson D J, and Comins J D 1999 *J. Appl. Phys.* **86** 1410-4

[8] Ribeiro D R S, Souza D N, Baldochi S L, and Caldas L V E 2008 *Radiation Measurements* **43** 1132-4

[9] Baldacchini G, Chiacchiaretta P, Gupta V, Kalinov V, and Voitovich A P 2008 *Physics of the Solid State* **50** 1747-1755

[10] Puchalska M and Bilski P 2006 *Radiation Measurements* **41** 659-664

[11] Vincenti M A, Montereali R M, Nichelatti E, Ambrico M, Schiavulli L, Kalinov V S, Voitovich A P, Baldacchini G, *ENEA Technical Report*, in press

[12] Vincenti M A, Baldacchini G, Kalinov V S, Montereali R M, Voitovich A P 2010 *IOP Conference Series: Materials Science and Engineering* **15** 012053

The mechanism of radiolysis of alkaline-earth nitrates

V Anan'ev[1], L Kriger[1] and M Miklin[2]

[1] Department of Analytical Chemistry, Kemerovo State University, Kemerovo, Russia
[2] Department of Physical Chemistry, Kemerovo State University, Kemerovo, Russia

e-mail: miklin@kemsu.ru

Abstract. The formation of peroxynitrite and nitrite in crystalline alkaline-earth nitrates under γ-irradiation at 310 K by optical reflectance spectroscopy has been studied. The radiolysis of $Sr(NO_3)_2$ and $Ba(NO_3)_2$ results in nitrite and peroxynitrite, $Mg(NO_3)_2{\cdot}6H_2O$ and $Ca(NO_3)_2{\cdot}4H_2O$ – nitrite. The mechanism for nitrite and peroxynitrite formation under γ-irradiation of crystalline alkaline-earth nitrates has been discussed.

1. Introduction

The radiolysis of solid inorganic nitrates has brought forth numerous experimental studies [1-6]. Gamma radiolytic decomposition of nitrates is known to produce nitrite – NO_2^-. Recently, it has been found out that it is peroxynitrite – $ONOO^-$ that is one of the radiolysis products of crystalline alkali nitrates [6-8]. The analysis of literature data shows that peroxynitrite is formed from high-energy excited state of the nitrate ion – NO_3^{-*} (> 4.8 eV) both in water solution [9] and in solids [6]. Since NO_3^{-*} can be formed under radiolysis of alkaline-earth nitrates it can be assumed that peroxynitrite can be formed too.

In this paper, the formation of peroxynitrite and nitrite under γ-irradiation of crystalline alkaline-earth nitrates by optical reflectance spectroscopy was studied to develop the mechanism of solid nitrates radiolysis.

2. Experimental

All nitrates used were analytical reagent grade. All solutions were prepared in doubly distilled water. The powders of barium, strontium, calcium, and magnesium nitrates were grown by slowly evaporating from the solutions. The presence of crystallization water in calcium and magnesium nitrate crystals was indicated by absorption in the ranges 3200-3900 and 1400-1800 cm^{-1} of the IR reflectance spectra. The formulas $Ca(NO_3)_2{\cdot}4H_2O$ and $Mg(NO_3)_2{\cdot}6H_2O$ were determined based on the water content in crystalline hydrates.

The samples were irradiated with ^{60}Co γ-rays at ~310 K. The dose rate 0.16 Gy/s was measured with a Fricke dosimeter, assuming the radiation chemical yield of Fe^{3+} to be equal to 15.6 $(100\ eV)^{-1}$. The dose absorbed by the sample was calculated using the mass energy absorption coefficients.

The optical reflectance spectra of both untreated and irradiated crystalline samples were recorded by means of a "Shimadzu UV-2450" spectrophotometer at room temperature. Figure 1 displays the optical reflectance spectra of untreated alkaline-earth nitrates. The band with maximum position at

293, 282, 271, and 275 nm for $Mg(NO_3)_2{\cdot}6H_2O$, $Ca(NO_3)_2{\cdot}4H_2O$, $Sr(NO_3)_2$, and $Ba(NO_3)_2$, respectively, was observed.

The infrared reflectance spectra of untreated samples were recorded at room temperature by means of a FSM 1201 spectrophotometer.

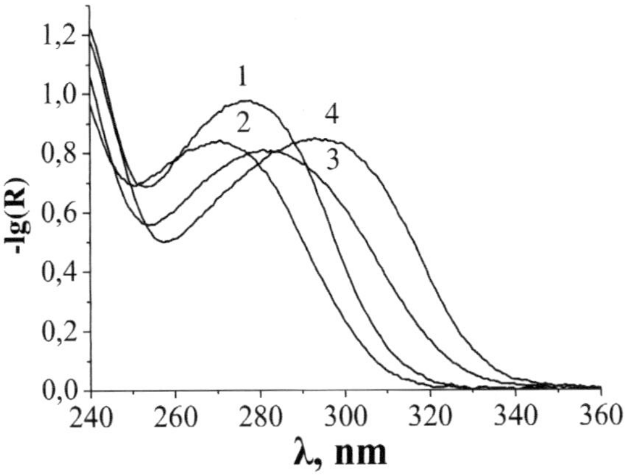

Figure 1. Optical reflectance spectra of crystalline $Ba(NO_3)_2$, $Sr(NO_3)_2$, $Mg(NO_3)_2{\cdot}6H_2O$ and $Ca(NO_3)_2{\cdot}4H_2O$ (1, 2, 3, and 4, respectively).

3. Results

Irradiation of the samples up to ~70 kGy results in new band in the optical reflectance spectra (Figures 2 and 3), whose maximum position at 344, 348, 335, and 338 nm for crystalline $Mg(NO_3)_2{\cdot}6H_2O$, $Ca(NO_3)_2{\cdot}4H_2O$, $Sr(NO_3)_2$, and $Ba(NO_3)_2$, respectively, is observed. The spectra of irradiated alkaline-earth nitrates are stable at room temperature. The maximum position of the broad band in γ-irradiated $Sr(NO_3)_2$ and $Ba(NO_3)_2$ are similar to the one of the optical absorption band due to peroxynitrite in these photolyzed crystals (333 and 340 nm, respectively) [10]. Based on this fact and the shape of the band it can be assumed that γ-induced band in the optical reflectance spectra due to peroxynitrite.

In contrast, the maximum position of the narrow band in γ-irradiated crystalline $Mg(NO_3)_2{\cdot}6H_2O$ and $Ca(NO_3)_2{\cdot}4H_2O$ are dissimilar to the broad band due to peroxynitrite in these photolyzed crystals (maximum position at 322 and 319 nm, respectively) [10]. Taking into account

i) the parameters of optical absorption band due to nitrite in alkali nitrates crystals doped with nitrite [8, 11];

ii) the parameters of optical absorption band due to nitrite in γ-irradiated $RbNO_3$ and $CsNO_3$ crystals [8];

iii) the details of the radiation-induced absorbance 320-380 nm with fine structure of at least nine peaks (see Table). They are similar to the fine structure of radiation-induced absorbance 320-380 nm, as detectable in irradiated $NaNO_3$ and KNO_3 crystals with absorbed doses 160 kGy [12] (see Table) due to species contains the O-N=O grouping,

we can conclude that radiation-induced band in γ-irradiated crystalline $Mg(NO_3)_2{\cdot}6H_2O$ and $Ca(NO_3)_2{\cdot}4H_2O$ is due to nitrite. It can be assumed that the observed band is a superposition of three individual bands which maximum position shifted relative to each other by 5-7 nm, because in the $NaNO_2$ crystal the vibration frequencies for the nitrite ion $\nu_1^{/}$ and $\nu_2^{//}$ in the B_1 state are equal to 1020 and 630 cm^{-1}, respectively [13].

Taking into account the fact that radiolysis of alkaline-earth nitrates results in nitrite, and based on our experimental results we can conclude that radiolysis of crystalline $Sr(NO_3)_2$ and $Ba(NO_3)_2$ results in nitrite and peroxynitrite, and crystalline $Mg(NO_3)_2{\cdot}6H_2O$ and $Ca(NO_3)_2{\cdot}4H_2O$ – nitrite.

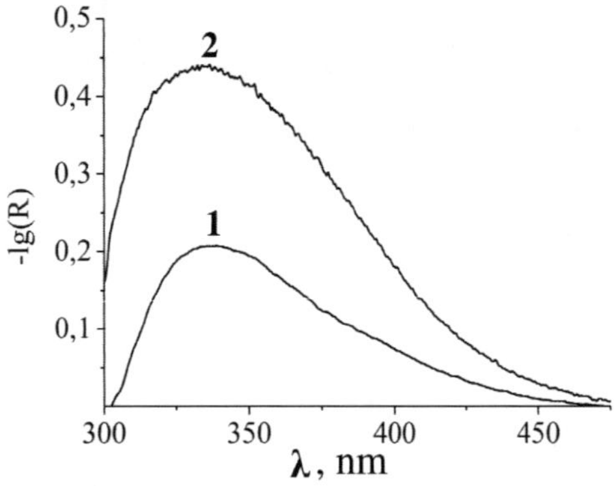

Figure 2. Gamma-induced optical reflectance spectra of γ-irradiated with dose 67.2 kGy crystalline $Ba(NO_3)_2$ (1) and $Sr(NO_3)_2$ (2).

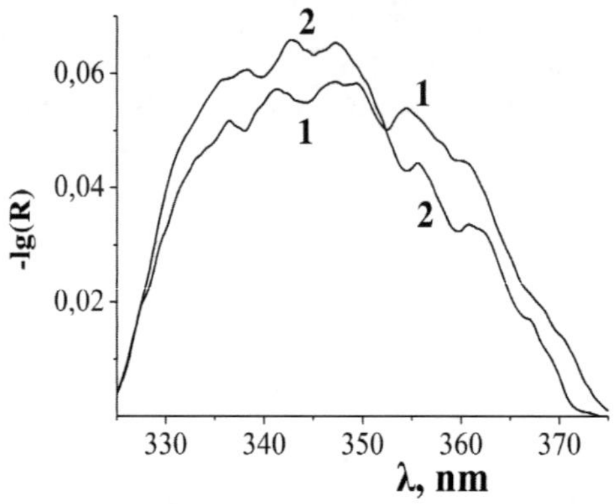

Figure 3. Gamma-induced optical reflectance spectra of γ-irradiated with dose 67.2 kGy crystalline $Mg(NO_3)_2{\cdot}6H_2O$ (1) and $Ca(NO_3)_2{\cdot}4H_2O$ (2).

4. Discussion

Since peroxynitrite, formed under photolysis of crystalline hydrated alkaline-earth nitrates, is stable [10], and it does not register in the optical reflectance spectra of γ-irradiated crystalline $Mg(NO_3)_2{\cdot}6H_2O$ and $Ca(NO_3)_2{\cdot}4H_2O$, we can conclude that the high-energy excited state of the nitrate ion, resulting in peroxynitrite formation, is not formed. A mechanism of the water effect on the radiolysis of crystalline hydrated alkaline-earth nitrates can be discussed based on the pathways of degradation of electronic excitations formed under irradiation.

Table. Fine-structure peak positions (nm) of optical absorption in γ-irradiated $NaNO_3$ and KNO_3 [12], optical reflectance absorption in irradiated $Mg(NO_3)_2{\cdot}6H_2O$ and $Ca(NO_3)_2{\cdot}4H_2O$.

$NaNO_3$	KNO_3	$Mg(NO_3)_2{\cdot}6H_2O$	$Ca(NO_3)_2{\cdot}4H_2O$
			338
343		342	344
349		349	349
355	356	358	356
362	360		362
370	372	369	369
378		375	374
	380	381	378
	388	386	387

Theoretical or experimental study of the energy bands structure of crystalline hydrated alkaline-earth nitrates is not available in the literature; therefore we consider the one for crystalline alkali nitrates. The structure of energy bands for them is characterized by the discontinuity of the spectrum of the vacant state [8] to result the first narrow conductivity band. The maxima in electron state densities corresponding to the first and the second conductivity bands are in the ~4 and ~10 eV regions, respectively, if the energy of the last occupied state is assumed to be zero. For alkaline-earth nitrate crystals the location of bands can be shifted as much as ~1 eV.

The initial stage of the radiolysis of crystalline alkali nitrates results in the charges with holes in valence band and electrons in the second conductivity band. Anan'ev and Seliverstov [8] suggested that an electron from this band can relax to the first conductivity band to generate ~ 6 eV with subsequent formation of a high-energy excited state of the nitrate ion. Triplet high-energy excited state of the nitrate ion can be dissociated to form the nitrite ion and singlet ones can be transformed to form the peroxynitrite ion. Electrons from the first conductivity band and the holes from the valence band can be recombined without dissociation. It can be assumed that this mechanism is suitable for anhydrous alkaline-earth nitrates based on the fact that under γ-irradiation nitrite and peroxynitrite are formed.

We assume, that in hydrated alkaline-earth nitrates, when an electron from the second conductivity band relax to the first conductivity band it generate energy sufficient for the ionisation of the water molecule to produce the complex [e^-, H_2O^+]. At the same time a high-energy excited state of the nitrate ion does not form.

The mechanism of formation of radiolysis product, taking into account the fact that reactions occur in "cage", is following. Electron, formed under the ionisation of the water molecule, reacts either with the nitrate ion ($k < 1\cdot10^{10}$ $M^{-1}c^{-1}$) to produce excited radical $(NO_3\bullet^{2-})^*$

$$NO_3^- + e^- \rightarrow (NO_3\bullet^{2-})^*$$

or with water molecule (hereinafter all rate constants are determined for water solution [14, 15])

$$e^- + H_2O \rightarrow e^- + OH\bullet + H^+$$

Electron fast react with H^+ ($k = 2.2\times10^{10}$ $M^{-1}s^{-1}$)

$$e^- + H^+ \rightarrow H\bullet$$

$OH\bullet$ react with the nitrate ion ($k < 1\cdot10^5$ $M^{-1}c^{-1}$) rather than $H\bullet$ ($k = 2.4$ $M^{-1}c^{-1}$)

$$NO_3^- + OH\bullet \rightarrow NO_3\bullet + OH^-$$

Formed under former reaction $H\bullet$ react with NO_3^- ($k = 2.4\times10^7$ $M^{-1}s^{-1}$)

$$NO_3^- + H\bullet \rightarrow HNO_3\bullet^-$$

Excited radical $(NO_3\bullet^{2-})^*$ either relax to ground state

$$(NO_3\bullet^{2-})^* \rightarrow NO_3\bullet^{2-}$$

or dissociate to produce the nitrite ion

$$(NO_3\bullet^{2-})^* + H_2O \rightarrow NO_2^- + OH\bullet + OH^-$$

Reaction of water molecule with $NO_3\bullet^{2-}$ ($k = 5.5\times10^4$ s^{-1}) results in the radical $NO_2\bullet$

$$NO_3\bullet^{2-} + H_2O \rightarrow NO_2\bullet + 2OH^-$$

$HNO_3\bullet^-$ may dissociated to produce the radical $NO_2\bullet$ too

$$HNO_3\bullet^- \rightarrow NO_2\bullet + OH^-$$

The reaction of both $NO_2\bullet$ and $NO_3\bullet$ with water results in the nitrite ion under dissolving of irradiated samples.

Thus, it is possible to consider the radiolysis of hydrated alkaline-earth nitrates as the one of water solution of the nitrate salts.

5. Conclusion

The mechanism of the radiolysis products formation in hydrated and anhydrous alkaline-earth nitrates can be accounted for by the peculiarities of degradation of electronic excitation. This mechanism allows us to explain the fact why radiation-chemical yield of nitrite in hydrated alkaline-earth nitrates is higher than that for anhydrous nitrates [16, 17]. This is due to

i) energy of irradiation is mainly spent on the formation of the complex [e^-, H_2O^+], but not on the formation of high-energy excited state of the nitrate ion;

ii) the value of rate constant for reaction of the nitrate ion with electron resulting in excited $(NO_3\bullet^{2-})^*$ formation is very fast.

Acknowledgements

This works was supported state assignment of Ministry of Education and Science of Russian Federation No 2014/64.

References

[1] Agrawal N, Garg A N 2004 *Radiat. Phys. Chem.* **69** 121
[2] Agrawal N, Garg A N 2005 *Radiat. Phys. Chem.* **73** 147
[3] Anan'ev V 2009 *Radiat. Phys. Chem.* **78** 394
[4] Johnson E R 1970 *The Radiation-induced decomposition of inorganic molecular ions.* Gordon&Breach Sci. Pub. N.-Y.-London-Paris
[5] Kulkarni S P, Garg A N 1986 *J. Radioanal. Nucl. Chem.* **98** 65
[6] Miklin M B, Nevostruev V A 1987 *Khimiya Vysokikh Ehnergij (Rus).* 21 154
[7] Anan'ev V, Kriger L, Poroshina M 2002 *Chem. Phys.Letters.* **365** 554
[8] Anan'ev V, Seliverstov M, 2005 *Ceramics – Silikaty* **49** 68
[9] Mack J, Bolton J R 1999 *J. Photochem. Photobiol. A* **128** 1
[10] Kriger L D, Miklin M B, Dyagileva E P, Anan'ev V A 2013 *Russian J. Phys. Chem. A* **87** 319
[11] Anan'ev V, Miklin M, Nevostruev V 2007 *J. Photochem. Photobiol. A* **186** 330
[12] Cunningham J 1964 *J. Chem. Phys.* **41** 3522
[13] Hangyo M, Yamanaka H, Kato R 1981 *J. Phys. Soc. Japan* **50** 895
[14] Gonzalez M G, Braun A M 1995 *Res. Chem. Intermed.* **21** 837
[15] Gonzalez M G, Oliveros E, Wörner M, Braun A M 2004 *J. Photochem. Photobiol.* C **5** 225
[16] Iyrgena T S, Bugaenko L T, Tiliks Y E 1981 *Khimiya Vysokikh Ehnergij (Rus).* **15** 422
[17] Parwate D V, Garg A N 1984 *J. Radioanal. Nucl. Chem.* **85** 203

IOP Conf. Series: Materials Science and Engineering **80** (2015) 012024 doi:10.1088/1757-899X/80/1/012024

Molecular dynamics simulations of spinels: $LiMn_2O_4$ and $Li_4Mn_5O_{12}$ at high temperatures

R.S. Ledwaba, M.G. Matshaba and P.E. Ngoepe
Materials Modelling Centre, University of Limpopo, Private Bag X 1106, Sovenga,0727, South Africa

E-mail: phuti.ngoepe@ul.ac.za

Abstract. Energy storage technologies are critical in addressing the global challenge of clean sustainable energy. Spinel lithium manganates have attracted attention due to their electrochemical properties and also as promising cathode materials for lithium-ion batteries. The current study focused on the effects of high temperatures on the materials, in order to understand the sustainability in cases where the battery heats up to high temperature and analysis of lithium diffusion aids in terms of intercalation host compatibility. It is also essential to understand the high temperature behaviour and lithium ion host capability of these materials in order to perform the armorphization and recrystalization of spinel nano-architectures. Molecular dynamics simulations carried out to predict high temperature behaviour of the spinel systems. The NVE ensemble was employed, in the range 300 - 3000K. The melting temperature, lithium-ion diffusion and structural behaviour were monitored in both supercell systems. $LiMn_2O_4$ indicated a diffusion rate that increased rapidly above 1500K, just before melting (~1700K) and reached its maximum diffusion at 2.756×10^{-7} cm^2s^{-1} before it decreased. $Li_4Mn_5O_{12}$ indicated an exponential increase above 700K reaching 8.303×10^{-7} cm^2s^{-1} at 2000K and allowing lithium intercalation even above its melting point of around 1300K. This indicated better structural stability of $Li_4Mn_5O_{12}$ and capability to host lithium ions at very high temperatures (up to 3000 K) compared to $LiMn_2O_4$.

1. Introduction

The spinel lithium manganese oxides (LMO) are some of the most promising alternative cathode materials for lithium-ion power batteries, identified by Thackeray et al [1]. This is due to their low cost, environmental compatibility, abundance, lower cost production demands, less toxicity and good thermal stability. Spinels show great promise due to their lithium-ion conductivity and three-dimensional structure, which gives them good structural stability [2]. The charged $LiMn_2O_4$ material shows higher thermal stability in the electrolyte at high temperature than recently commercialised $LiCoO_2$ and $LiNiO_2$. Spinel $Li_4Mn_5O_{12}$ exhibits good electrochemical stability at 3 V and has better cyclability than spinel $LiMn_2O_4$ [3, 4]. This material is comparable to $LiMn_2O_4$ and the two have the same spinel structure. $Li_4Mn_5O_{12}$ has a good reversibility, which has attracted great interest due to the absence of Jahn–Teller distortion, unlike $LiMn_2O_4$ [5]. Spinel $LiMn_2O_4$ and $Li_4Mn_5O_{12}$ offer theoretical capacities of 148 mAh/g [3] and 163 mAh/g, respectively in 3V region [4]. The edge-shared octahedral $LiMn_2O_4$ host structure is highly stable, possesses a series of intersecting tunnels.

The tunnels allow for three-dimensional diffusion of lithium ions [5, 6] into/from the 8a tetrahedral sites and maintain the initial cubic spinel symmetry. However, Jahn-Teller distortion causes severe damage to spinel $LiMn_2O_4$ cathode and affects the three-dimensional Li diffusion pathway, which results in the capacity fade of active material. $Li_4Mn_5O_{12}$ has Mn oxidation state (+4), which suppresses the Jahn Teller distortion [5, 6] and nano-architectures can also be used to reduce electron and Li-ion path. They can also enhance capacity rate and cycling stability. Previous research indicated the calcination temperature, working voltage, current density, electrolyte species and scan rate had great effect on the electrochemical performance of $Li_4Mn_5O_{12}$ material [6]. The current study will focus on the effect of high temperatures on such rechargeable battery materials (cathodes).

2. Computational Methods

2.1. Molecular Dynamics

The energy minimization code (METADISE) [7] was used to generate supercell systems and DL_POLY input files required for simulations. All molecular dynamics simulations were carried out using the DL_POLY computer code [8], on systems with 12096 atoms for $LiMn_2O_4$ and $Li_4Mn_5O_{12}$. The calculations were run for 250ps, and simulation temperatures were in the range 300 - 3000K for all spinel systems. All calculations presented in this study are based on the Born model of the ionic solid, in which the Mn^{4+}, Mn^{3+}, Li^+ and O^{2-} ions interact via long-range Coulombic interactions and short-range parameterized interactions and the corresponding potential parameters have been reported previously [9]. The electrostatic forces were calculated using Ewald sum [10] with automatic parameter optimisation.

2.2. The Spinel Systems

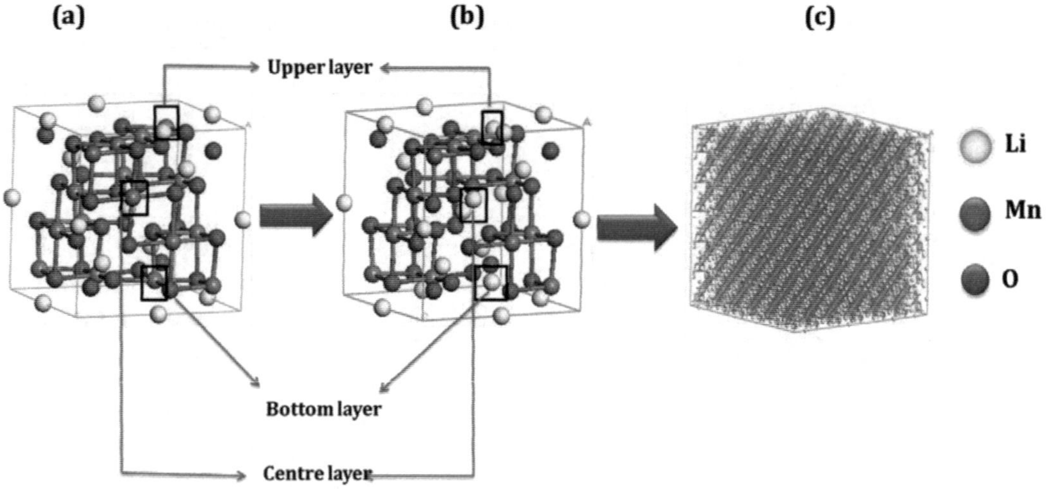

Figure 1: (a) Spinel $LiMn_2O_4$ has a = 8.24 Å [7] and (b) $Li_4Mn_5O_{12}$has a = 8.149 Å [8], space group Fd-3m and FCC lattice. The supercell system (c) consists of 12096 atoms

3. Results and discussion

Figure 2 illustrates the spinel $LiMn_2O_4$ structures and their corresponding radial distribution functions (rdfs) at 300 K, 1700 K and 3000 K. The structure at 300K is in a crystalline form and its rdfs at 300K depict sharp peaks with high probability values. These peaks imply that the atoms are arranged in such a way that they can be found at a certain distance (r) in relation to the reference atom. At 1700 K, the appearance of the atoms in the system is completely different from the lower temperature arrangements. This amorphous state of the system indicates that the structure has reached a melting point and this change in phase results in random motion of ions in the system. The rdf peaks, for 3000 K, are very broad peaks with reduced probability of location compared to the lower temperature ones, and indicate very weak bonds between the atoms. This analysis indicates that the melting point of spinel $LiMn_2O_4$ is around 1700 K and the structure transforms to liquid phase above that temperature. Figure 3 illustrates the rdfs and structures of spinel $Li_4Mn_5O_{12}$ at different temperatures. A similar trend is observed where the structure shows extensive disordering above 1250 K after the material had melted and the rdf peaks are also very broad and fewer compared to the system at 300 K. Figure 4 below illustrates variation of ion diffusion in spinel manganates at different temperatures. Spinel $LiMn_2O_4$ shows lithium diffusion above 500K and a rapid increase is observed above 1000 K until it is reversed at 2000 K with a maximum diffusion of 2.756 x 10^{-7} cm^2s^{-1} . A similar initial trend is observed for $Li_4Mn_5O_{12}$ but it shows a rapid increase after the sintering temperature and allows for lithium ion diffusion at very high temperatures reaching $8.303x10^{-7}$ cm^2s^{-1} at 2000K. This shows that $Li_4Mn_5O_{12}$ is a more compatible cathode material candidate than $LiMn_2O_4$. This high temperature study will guide attempts of simulating spinel $LiMn_2O_4$ nano-architectures by amorphisation recrystallization methods [9].

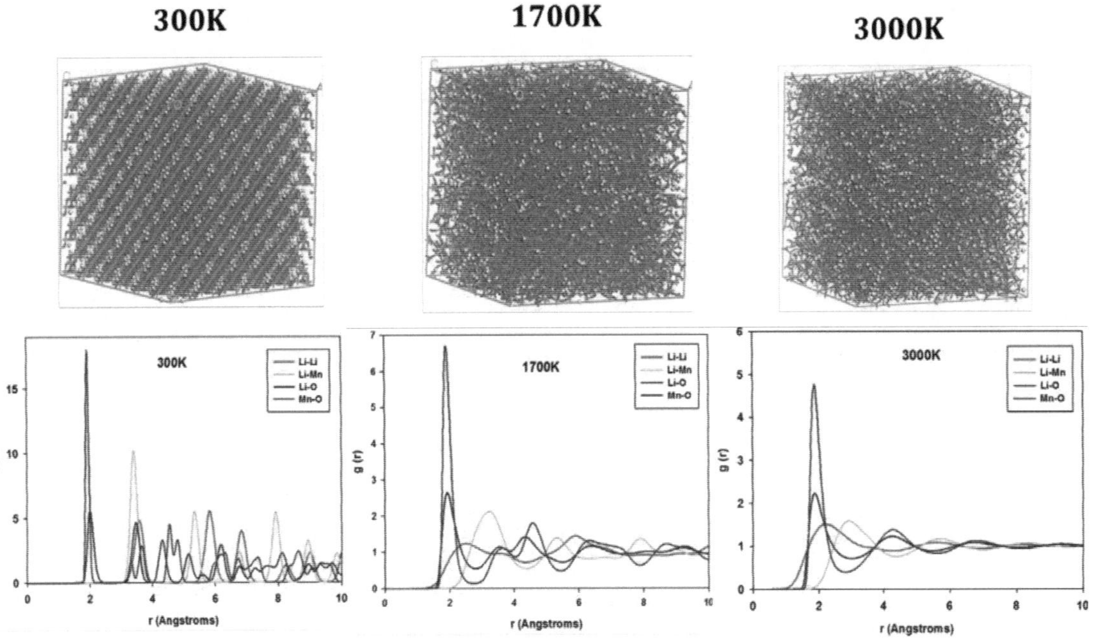

Figure 2: Structural snapshots and rdf plots of spinel $LiMn_2O_4$ supercell at: 300, 1700 and 3000K. Li-Li=red, Li-Mn=green, Li-O=blue, Mn-O=purple.

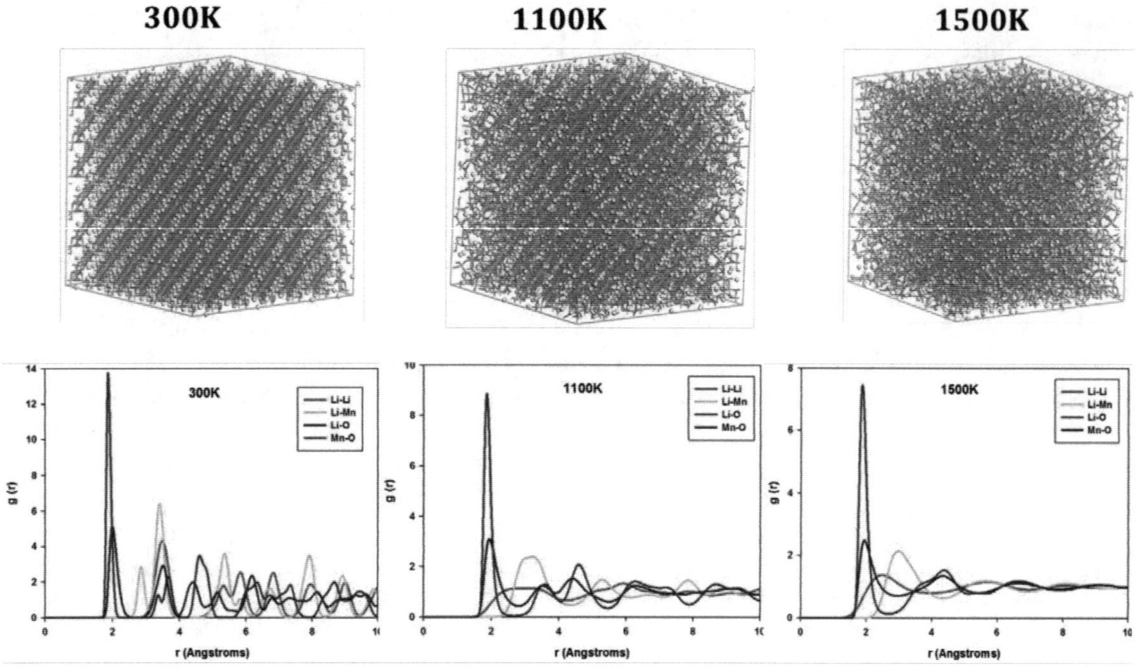

Figure 2: Structural snapshots and rdf plots of spinel $Li_4Mn_5O_{12}$ supercell at: 300, 1100 and 1500K. Lines colours as in Figure 2.

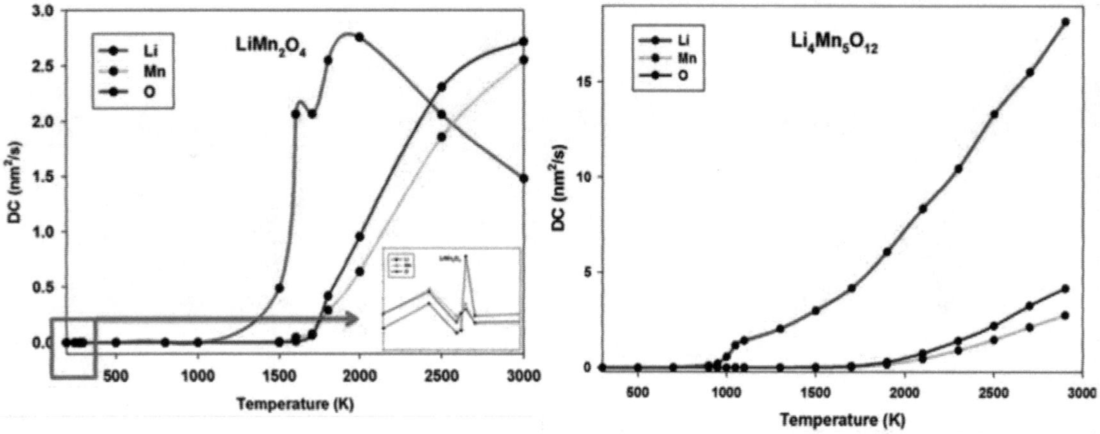

Figure 3: Diffusion coefficients for Li, Mn and O ions in spinel $LiMn_2O_4$ and $Li_4Mn_5O_{12}$. Li=red, Mn=green, O=purple.

4. Conclusion

The study has shown different high temperature effects in the spinel manganese oxide systems. Melting points of the two systems have also being observed to be approximately 1700K for $LiMn_2O_4$ and 1250 K for $Li_4Mn_5O_{12}$. The melting temperature region shows a decrease in lithium ion mobility for $LiMn_2O_4$ and an exponential increase in the lithium ion mobility for $Li_4Mn_5O_{12}$. This study has shed valuable insights on high temperature behaviour of the two types of investigated spinels. It will also inform a complex process of generating spinel nano-architectures by amorphization recrystallization method.

Acknowledgements
We acknowledge support of the South African Research Chair Initiative of the Department of Science and Technology and the National Research Foundation in Pretoria and the Centre for High Performance Computing in Cape Town.

References

[1] Voelcker J., *IEEE Spectrum,* **44**, 26 (2007)
[2] Thackeray M. M., David W. I. F., Bruce P. G., and Goodenough J. B., *Materials Research Bulletin,* **18**, 461 (1983)
[3] Tanaka Y., Zhang Q. and Saito F., *Powder Technology,* **132**, 74 (2003)
[4] Piszora P., Nowicki W., Darul J. , Carlson S. and Cerenius Y. , *Synchrotron Radiation in Natural Science,* **7**, 62 (2008)
[5] Zhao Y., Lai Q., Zeng H., Hao Y. and Lin Z. , *Springer,* DOI 10.1007/s11581-013-0860-y (2013)
[6] Hao Y.J., Wang Y.Y., Lai Q.Y., Zhao Y., Chen L.M.and Ji X.Y., *Journal of Solid State Electrochemistry,* **13**, 905 (2009)
[7] Watson G.W., Kelsey E. T., De Leeuw N. H., Harris D. J. and Parker S. C., *Journal of the Chemical Society,* **92**, 433 (1996)
[8] Smith W. and Forester T. R., *Journal of Molecular Graphics,* **14**, 136 (1996)
[9] Maphanga R. R., Sayle T. X. T., Ngoepe P. E. and Sayle D. C, *Physical Chemistry Chemical Physics,* **13**, 1307 (2011)
[10] Ewald P.P., *Annals of Physics,* **64**, 253 (1921)

IOP Conf. Series: Materials Science and Engineering **80** (2015) 012025 doi:10.1088/1757-899X/80/1/012025

First-principles study of electronic structure of Ce^{3+} centres in alkaline-earth fluorides including spin-orbit and scalar relativistic effects

N.V. Popov[1,2], E.A.Radzhabov[2], A.S.Mysovsky[1,2]

1 - National Research Irkutsk State Technical University, 83 Lermontov street, Irkutsk, Russia
2 - A.P. Vinogradov Institute of Geochemistry SB RAS,1a Favorsky St., Irkutsk, Russia

Email:nvpopov@gmail.com, eradzh@igc.irk.ru, mysovsky@gmail.com

Abstract. In this paper we present the results of *ab initio* calculations of Ce^{3+} impurity centres in CaF_2 crystal. The cubic and tetragonal (interstitial fluorine compensated) configurations of centres were considered. The embedded cluster approach with polarizible crystalline lattice and CASSCF/CASPT2 methods were used to study the structural configurations and electronic properties of defects under consideration. Scalar relativistic corrections were taken into account at Douglas-Kroll-Hess level and spin-orbit effects were calculated with the RASSI approach. Calculated optical absorption spectra for cubic and tetragonal centres demonstrate good agreement with the experiment. Small symmetry breaking displacements of neighbouring ions were observed for cubic Ce^{3+} centre ground state.

1. Introduction

Cerium-doped compounds have received considerable interest due to their wide applications in scintillators and solid state lasers. Well established experimental and theoretical data for cerium impurity in alkaline-earth fluoride crystals are to be found in Refs. [1, 2]. In these crystals Ce^{3+} ion is supposed to form centres with cubic symmetry (for example, such centres are considered in Ref. [3]) as well as tetragonal and trigonal centres with charge compensating fluorine interstitial [2]. However, EPR spectroscopy does not confirm the existence of cubic centres [4, 5] although they can be readily observed by the means of optical and infrared spectroscopy, which are less sensitive to the defect symmetry then EPR technique.

The purpose of the present theoretical study is to build adequate model of spatial and electronic structure of Ce3+ centres in CaF_2 crystals including scalar relativistic effects and spin-orbit interaction which strongly affects energy levels of rare earth impurities.

2. Calculation details

Embedded cluster approach used in this work allows one to divide the crystal with the defect into several regions described at different levels of theory: 1) quantum-mechanical (QM) cluster which is calculated *ab initio*; 2) classical region where the interaction between ions is described with pair potentials; 3) region of fixed point charges reproducing proper crystalline field inside 1+2 region. Geometry optimization is allowed for both regions 1 and 2. For quantum chemical calculations we

used the Molcas package [6]. ANO-RCC all electron basis set [7, 8] was used for Ce, F and Ca ions inside QM cluster. In the classical region *ab initio* model potentials (AIMPs) [9] were placed on Ca and F ions in order to avoid possible distortion of QM cluster electronic density caused by the presence of bare Coulomb potentials. Geometry optimization was performed at the complete active space self consistent field (CASSCF) level. State averaged CASSCF calculations with active space containing 4f, 5d and 6s orbitals of Ce ion were performed for treating ground and excited electronic states of the defects. Correlation was taken into account with complete active space second order perturbation theory (CASPT2) method. To calculate spin-orbit coupling we used restricted active space state interaction (RASSI) approach.
We have studied two possible configurations of Ce^{3+} centres: the cubic and the tetragonal one with nearest interstitial fluorine ion. Quantum clusters $(CeF_8)^{5-}$ (Fig. 1) and Ca_5CeF_{13} (Fig. 2) were used for O_h and C_{4v} centres, respectively.

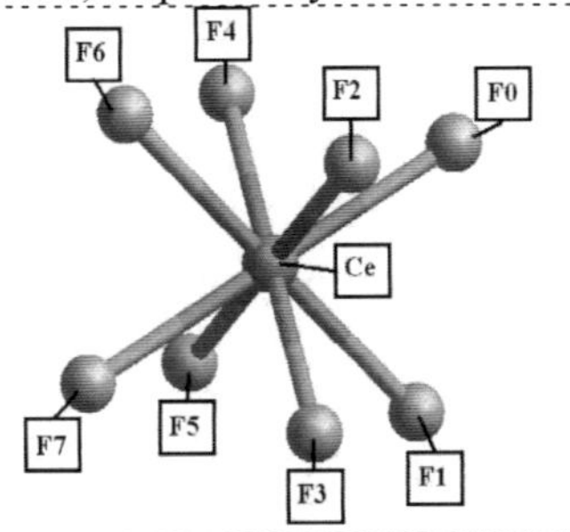

Fig. 1. CeF_8 quantum cluster containing Ce^{3+} ion in cubic configuration

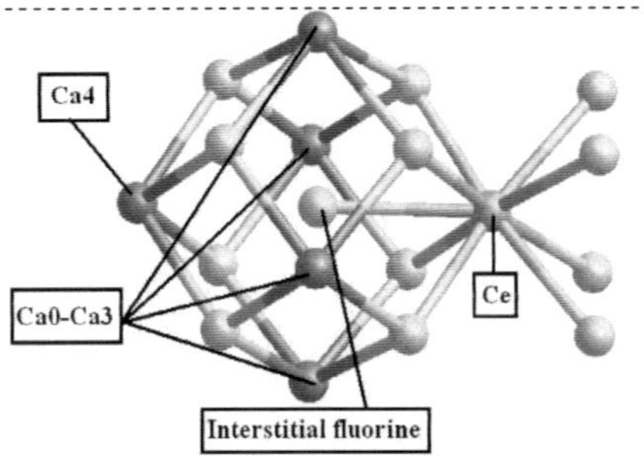

Fig. 2. Ca_5CeF_{13} quantum cluster containing Ce^{3+} and charge compensating interstitial F^- ion in tetragonal configuration

3. Result and discussion

After the geometry optimization of the ground state of the cubic Ce^{3+} centre the symmetry lowering from cubic to trigonal was observed. The symmetry unique displacements are shown in Table 1. The third order symmetry axis is the cube diagonal connecting fluorine ions F0 and F7 as they are denoted in Fig. 1. However, one might see that the displacements breaking O_h symmetry are very small – about 0.006 Å. Such a small value of displacements does not allow us to conclude that the symmetry of so called "cubic" Ce^{3+} centre does indeed break due to Jahn-Teller effect. On the other hand, there is no EPR evidence of the existence of Ce^{3+} centres with cubic symmetry [4,5]. The possible reasons of such behaviour remain unclear and might include dynamic Jahn-Teller effect as well as the presence of other impurities in the vicinity of Ce^{3+} centre.
For the tetragonal Ce^{3+} centre with nearest charge compensating interstitial fluorine ion the results of geometry optimization are shown in Table 2. The notation of atoms in Table 2 is the same as on Fig. 2. In both Tables 1 and 2 X, Y, Z are Cartesian coordinates of the atoms before the geometry optimization and ΔX, ΔY, ΔZ are the resulting displacements. Only symmetry unique atoms are listed. Tables 3 through 6 contain the data on energy levels and optical transitions for the cubic and tetragonal Ce^{3+} centres calculated with multistate CASPT2 method. The 4f-5d optical transitions and absorption spectrum of the cubic centre are shown in Table 4 and on Fig. 2a. The same information for tetragonal centre is represented in Table 6s and Fig. 2b. The calculated optical absorption spectra for both centres demonstrate good agreement with experimental [2,10] and theoretical [1,11,12] works.
According to Ref. [10] the high energy part absorption spectrum of cubic Ce^{3+} centre contains three distinct lines at about 53000 cm^{-1} split by 1700 and 1900 cm^{-1}. It is important to note that in purely cubic O_h coordination the first two of them must be degenerate. In our calculations due to small symmetry breaking displacements three different lines were obtained, but the separation between the

first and the second one is only about 350 cm^{-1}. This possibly indicates that asymmetric distortion of the cubic Ce^{3+} centre should be larger than we obtained.

Concerning the tetragonal Ce^{3+} centre one can see that the representative peak experimentally observed at about 40000 cm^{-1} which is due to the interaction between Ce^{3+} and interstitial fluorine ion is present in the calculated spectrum with both energy and oscillator strength being in good agreement with experiment.

Table 1. Structural parameters of the cubic Ce^{3+} centre.

Atom	X, Å	Y, Å	Z, Å	ΔX, Å	ΔY, Å	ΔZ, Å
Ce	0.00	0.00	0.00	0	0	0
F0	1.38	1.38	1.38	0	0	0
F1	1.38	1.38	-1.38	0.004	0.004	0
F6	-1.38	-1.38	1.38	-0.004	-0.004	0
F7	-1.38	-1.38	-1.38	0	0	0

Table 2. Structural parameters of the tetragonal Ce^{3+} centre

Atom	X, Å	Y, Å	Z, Å	ΔX, Å	ΔY, Å	ΔZ, Å
Ce	1.38	0	0	-0.12	0	0.03
Ca0-Ca3	-1.38	0	2.76	-0.01	0	0.10
Ca4	-4.14	0	0	0.10	0	0
F_i	-1.38	0	0	0.05	0	0

Table 3. 4f energy levels of the cubic Ce^{3+} centre

State	Energy without spin-orbit, cm^{-1}	Energy with spin-orbit, cm^{-1}
1	0.0	0.0
2	18.8	5.0
3	24.7	507.3
4	90.0	2250.5
5	91.0	2272.2
6	94.0	2290.0
7	2406.6	4064.8

Table 4. 5d energy levels and optical transitions of the cubic Ce^{3+} centre

State	Energy without spin-orbit, cm^{-1}	Energy with spin-orbit, cm^{-1}	Energy,cm^{-1}, (calc.) [11]	Energy, cm^{-1}, (expt.) [10]	Oscillator strength
8	31888	33058.9	33633	32300	1.35E-02
9	31892	33072.7		32300	6.8E-03
10	56442	57190.2	48071	51600	9.12E-03
11	56444	57539.5		53300	4.65E-03
12	56708	58909.2		55200	1.35E-03

Table 5. 4f energy levels of the tetragonal Ce^{3+} centre

State	Energy without spin-orbit, cm^{-1}	Energy with spin-orbit, cm^{-1}	Energy, cm^{-1} (calc.) [12]	Energy, cm^{-1} (expt.) [2]
1	0	0	0	0
2	97.3	275.2	281	110
3	98.4	512.5	559	579
4	180.9	2314.7	2276	2192
5	188.9	2316.7	2380	2307
6	636.3	2679.6	2635	2440
7	1892.0	3630.6	3325	3562

Table 6. 5d energy levels and optical transitions of the tetragonal Ce^{3+} centre

State	Energy without spin-orbit, cm^{-1}	Energy with spin-orbit, cm^{-1}	Energy, cm^{-1} (calc.) [11]	Energy, cm^{-1} (calc.) [12]	Energy, cm^{-1} (expt.) [2]	Oscillator strength
8	32692.8	33852.9	32907	30714	32350	1.26E-02
9	38591.3	39708.4	41457	40285		8.28E-04
10	51975.3	52771.9	46377	48992	49490	2.59E-03
11	51975.9	53539.9		49961	51160	5.20E-03
12	53467.3	54997.4	47506	52946	53370	4.80E-03

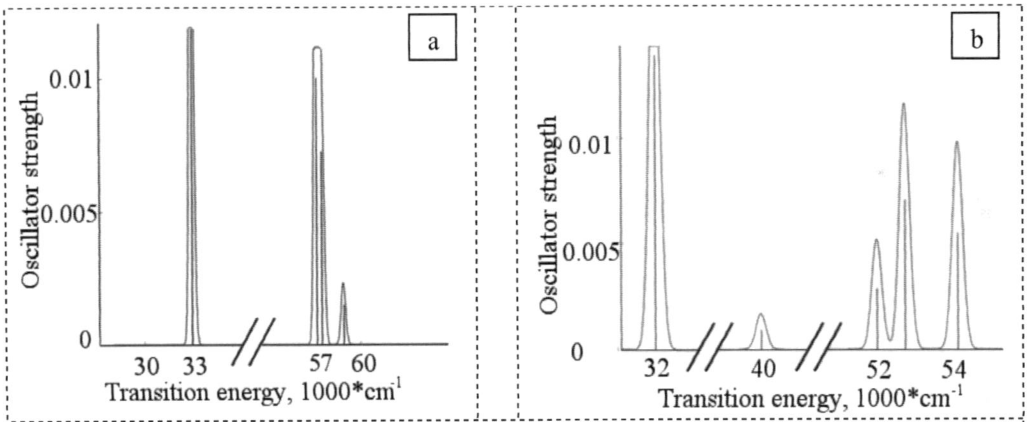

Fig. 2. Calculated optical absorption spectra of a) cubic and b) tetragonal Ce^{3+} centre

4. Conclusion

We performed *ab initio* embedded cluster calculations of Ce^{3+} impurity centres in CaF_2 crystal including scalar relativistic effects and spin-orbit coupling. Cubic and tetragonal centre were investigated. Calculated structural parameters, energy levels and absorption spectrum are in good agreement with experimental and theoretical data. This shows the applicability of the chosen approach for studying rare earth impurities in fluoride crystals.

The “cubic” Ce^{3+} centre turns out not to be perfectly cubic. We obtained small but not negligible asymmetric displacement of surrounding fluorine ions in the ground state of the centre. The deviation of this defect from perfect O_h symmetry can be seen from EPR data and from the splitting of high energy lines in optical absorption spectrum. At present moment it is not clear whether the symmetry breaking of Ce^{3+} centre should be attributed to static or dynamic Jahn-Teller effect or to the presence of other charge compensating impurities in the vicinity of the defect.

References

[1] Pack D W *et al* 1989 *Phys. Rev. B* **40** 9930
[2] Manthey W J 1973 *Phys. Rev. B* **8** 4086
[3] Barandiarán Z and Seijo L 2006 *Theor. Chem. Acc.* **116** 505
[4] Weber M J and Bierig R W 1964 *Phys. Rev.* **134** A1492
[5] Yamaga M *et al* 2004 *J. Lumin.* **108** 307
[6] Karlström G *et al* 2003 *Comput. Mater. Sci.* **28** 222
[7] Roos B O *et al* 2008 *J. Phys. Chem. A* **112** 11431
[8] Roos B O, Veryazov V and Widmark P O 2003 *Theor. Chem. Acc.* **111** 345
[9] Pascual J L and Seijo L 1995 *J. Chem. Phys.* **102** 5368
[10] van Pieterson L *at el* 2002 *Phys. Rev. B* **65** 045113
[11] Myasnikova A, Mysovsky A and Radzhabov E 2012 *IEEE Trans. Nucl. Sci.* **59** 2065
[12] Ning L *et al* 2012 *J. Phys. Chem. C* **116** 18419

Lanthanum-gallium tantalate: Heterogeneity and point defects

N S Kozlova[1], O A Buzanov[2], I S Didenko[1], A P Kozlova[1, 3], N A Siminel[1] and E V Zabelina[1]

[1] National University of Science and Technology "MISIS", Moscow, Russia
[2] Company Fomos-Materials, Moscow, Russia
E-mail: anna_kozlova_2009@mail.ru

Abstract. In this study the important factors having influence on the occurrence of color centers (crystal growth conditions and postgrowth effect on the crystal) in LGT ($La_3Ga_{5,5}Ta_{0,5}O_{14}$, langatate) crystals were investigated. In order to check the model of defect formation in crystals of langatate the effect of ionizing radiation on the spectral dependence of the attenuation coefficients of langatate crystals was studied. The ionizing radiation doses were 10^{11} – 10^{15} cm^{-2}. The optical homogeneity was evaluated by interferometric method on a Fizeau interferometer. The refractive indices of the crystals obtained in different atmospheres and after the isothermal annealing were measured by the prism method.

1. Introduction

Lanthanum - gallium tantalate ($La_3Ga_{5,5}Ta_{0,5}O_{14}$, langatate, LGT) is promising piezoelectric material [1]. It belongs to a class of symmetry 32 and is one of the isomorphic compounds of langasite ($La_3Ga_5SiO_{14}$, langasite, LGS), has a unique combination of physical properties, making it one of the most promising materials for use in pressure sensors.

Peculiarity of langatate is structure complexity of unit cell and a tendency to disordered distribution matrix ions into sites of crystal lattice [2, 3]. Irregularity and inhomogeneity of obtained crystals lead to heterogeneity of physical properties. Determination of growth conditions of homogeneous crystal allows to applying them more effective and improving the technology.

One of the major issues in the study of LGT is their color, which is determined by the presence of color centers. The important factors having influence on the occurrence of color centers are crystal growth conditions and post-growth treatment on the crystal. For this purpose methods of optical spectroscopy and electron irradiation were used.

2. Experimental procedure

In this study, we performed experiments with langatate crystals produced by the company Fomos-Materials. The crystals were grown in iridium crucibles in atmospheres of argon and in a mixture of argon and oxygen (Ar +(2%)O_2) and (Ar +(<2%) O_2) and in air, in platinum crucible. The samples represented plates of polar cut polished on both sides. To assess the quality crystal samples of langatate were studied by interferometry on a Fizeau interferometer. The samples were irradiated by different electron doses and then transmission spectra were measured. The irradiation doses were varied from $1\cdot10^{11}$ to $3\cdot10^{15} cm^{-2}$. A controllable linear accelerator (LU-6) as the source of mono-energetic (6 MeV) electrons was used in experiment, where a pulse (5 - μs) electron irradiation mode was realized. The samples of 2 - mm thickness were oriented perpendicular to the electron beam. The transmission spectra were measured with a Cary – 5000 UV-VIS-NIR (Varian) spectrophotometer within the spectral range 180 – 3300 nm. The transmission spectra measurements were performed before irradiation and after each dose of irradiation.

[3] Author to whom any correspondence should be addressed

3. Results and discussion

Langatate crystal samples were studied by interferometry in accordance with the scheme shown in figure 1. In this case, it was observed overlapping interference patterns formed by different types of bands.

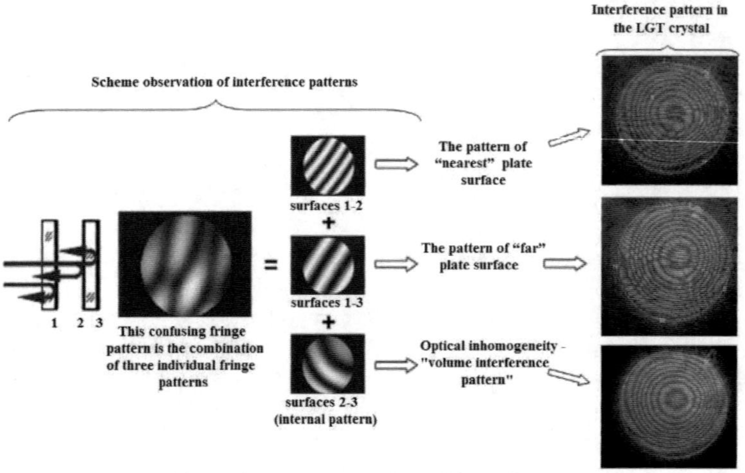

Figure 1. Scheme observation of interference patterns and interference pattern in the LGT crystal: 1 – the lower working surface of the internal wedge Fizeau plate, 2, 3 – the opposite sides of the sample.

Bands of the first type – frequent bands of different curvature – arise due to deviations from flatness of the plate surface. In this case, the interference is observed between the working surface of the wedge-shaped plate (1) and the investigated surface of the sample (2 or 3). Such patterns are different for the opposite surfaces of the plate and disappear when wedge-shaped plate removes from the interferometer. Bands of the second type form concentric circles. Their appearance depends neither on whether the observed plate surface is the upper nor the presence or absence of the wedge plate. The causes of these bands were studied. The phenomenon of interference is described by the equation (1):

$$\delta = 2 \cdot h \cdot n \cdot \cos r + \frac{\lambda}{2} \qquad (1)$$

where δ - path-length difference, h – the thickness of the plate, n – refractive index, r – angle of incidence, λ - wavelength.

In accordance with the equation (1), the path-length difference depends on the plate material inhomogeneity: the thickness unevenness of the plate, change the angle of incidence of light or the refractive index. Surface morphology was complex and did not correlate with the shape of the interference pattern in the plate bulk. Consequently, the cause of this pattern is the heterogeneity of the refractive indices.

Therefore, direct measurements of the refractive indices were carried out by the prism method on a goniometer-spectrometer GS-2. The radiation source was a helium lamp. For measurements we used the langatate prism with the base surfaces perpendicular to the optical axis of the crystal. It allows to measure the main refractive indices No and Ne. The refractive indices were measured at a wavelength of λ = 587,5 nm in different areas along the prism with a step of ~ 2 mm.

Analysis of the results shows that the values of the refractive indices of the crystals grown in different atmospheres of Ar and Ar + (2%) O_2 are heterogeneous along the length of the samples and change in the third decimal place (table 1).

Table 1. The refractive indices No and Ne of the crystals grown in different atmospheres

Growth atmosphere	refractive indices	
	No	Ne
Ar	1,9447 ÷ 1,9468	1,9723 ÷ 1,9745
Ar + (2%) O_2	1,9439 ÷ 1,9476	1,9715 ÷ 1,9746

The refractive indices decrease from the center towards the periphery of the crystal, independent of the growth atmosphere. Heterogeneity of the refractive index No for the crystal obtained in an atmosphere of Ar + (2%)O_2 is higher than in the crystals obtained in an atmosphere of Ar. This indicates a higher optical homogeneity of the crystals grown in argon atmosphere. Similar results were obtained for the refractive index Ne.

The important condition in technology of growth of langatate is the choice of growth atmosphere. The color intensity decreases from the crystal obtained in argon with 2% of oxygen (it is bright orange) to the crystal obtained in argon or in air (it is almost transparent and colorless) [3]. Transmission spectra of crystals obtained under the different conditions in initial state (before irradiation) are presented in figure 2.

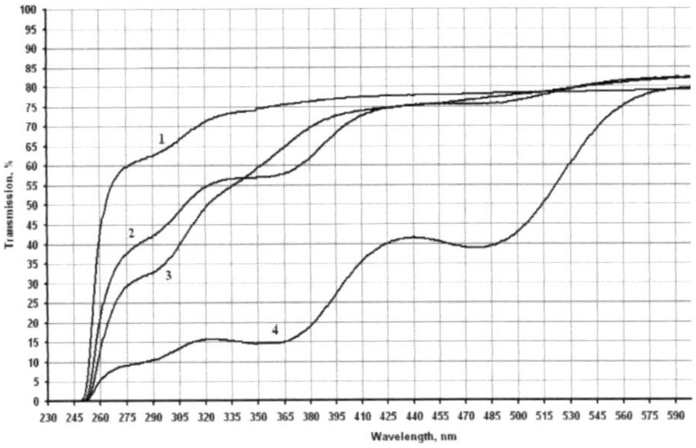

Figure 2. Transmission spectra of crystals obtained under the different conditions (atmospheres 1 - in (Ar), 2 - (Ar +(<2%) O_2), 3 – air, 4 - (Ar +(2%)O_2)) in initial state [3].

As the most changes of transmission spectra of the crystals are observed in the visible range transmission spectra are shown in wavelengths from 230 to 590 nm. Absorption bands are observed in all examined crystals regardless of growth atmosphere at wavelengths 290, 360, 480 nm. The absorption bands indicate presence of several point defects or their complexes in the form of color centers.

It is obvious, that the reason of crystal coloration is the presence of color centers. But a pattern of defect formation in langatate remains unknown. Some authors attribute [5 – 7] langatate coloration to different concentration of oxygen vacancies. Combination oxygen vacancy (V_o^{++}) and two electrons corresponds to F-center. It is logical that maximum concentration of oxygen vacancies is observed in crystal obtained in Ar, minimum concentration in the air (as shown in equation (2)). In this case according to the offered pattern crystal obtained in atmosphere of Ar must be most colored. So it is contrary to real coloration and spectra (Figure 2).

$$\left[V_o^{++}{}_{(air)}\right] < \left[V_o^{++}{}_{(Ar+(2\%)O_2)}\right] < \left[V_o^{++}{}_{(Ar+(<2\%)O_2)}\right] < \left[V_o^{++}{}_{(Ar)}\right] \quad (2)$$

Considerable changes of defect structure were the result of irradiation in crystal obtained in Ar atmosphere (Figure 3a). The absorption bands at 290 and 360 nm appeared even at the minimum dose irradiation. Absorption intensity increases more than twice. It means color centers at these wavelengths are unstable. Increasing of irradiation dose leads to intensity changes from 5 to 10% for absorption bands at 420 - 480 nm and from 15 to 35% for absorption bands at 290 - 360 nm.

Quite the contrary influence of irradiation on color centers of crystal obtained in (Ar+(~2%)O_2) atmosphere is insignificant (Figure 3b). Color centers generated in process of growth are stable enough.

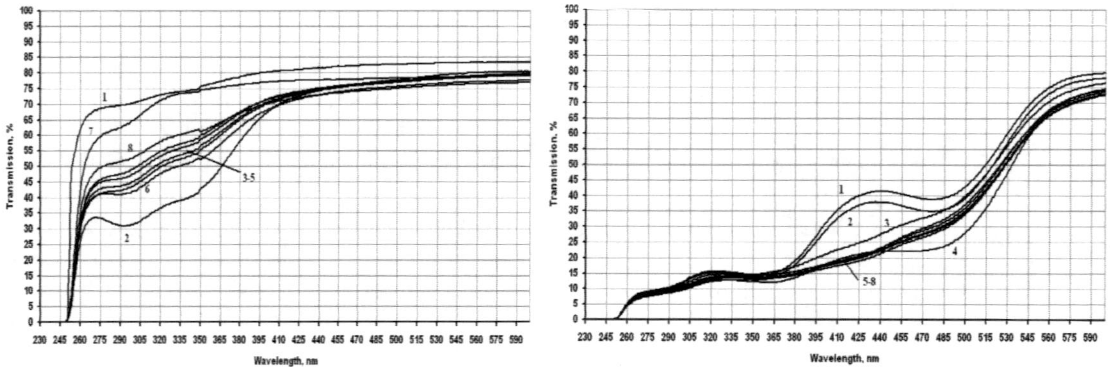

a b

Figure 3. Transmission spectra of crystals obtained in Ar (a) and Ar+(2%)O_2 (b) after different irradiation doses (1-initial state, 2-8 – increasing electron doses)

Increase of irradiation dose has an essential effect on absorption intensity at 420-480 nm for langatate obtained in Ar atmosphere and at 290-360 nm for langatate obtained in atmosphere (Ar+(~2%)O_2), see figures 4. Strong changes are observed at dose $1 \cdot 10^{12}cm^{-2}$.

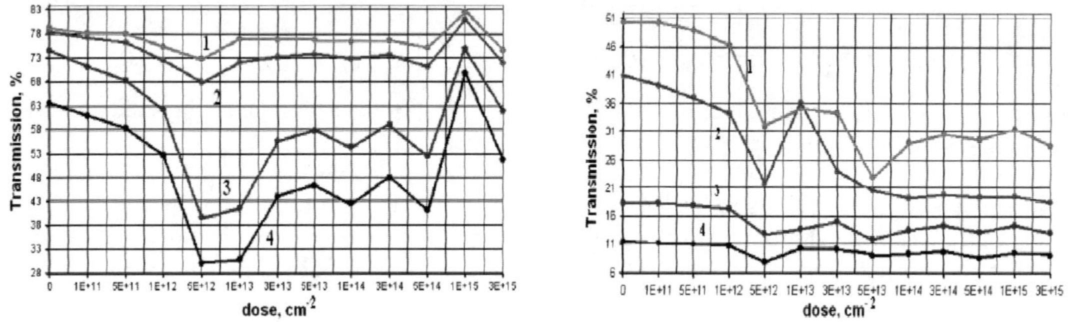

a b

Figure 4. Dose dependence for langatate samples grown in Ar (a) and Ar+(2%)O_2 (b) atmospheres at different wavelength: 1 - 480 nm, 2 - 420 nm, 3 - 360 nm, 4 - 290 nm.

These crystals were additionally investigated by X-ray diffraction analysis before and after irradiation. The results are evidence of vacancy mechanism of defect formation in langatate crystals. Vacancy concentration increases after irradiation. Lattice distortion takes place. Values of the spacing parameter "a" is shown in the table 2. Spacing parameter in crystal obtained in Ar+(2%)O_2 is much less than in crystal grown in Ar and both differ from tabulated value 8.2345 Å. Irradiation leads to decrease of value of spacing parameter for langatate obtained in Ar atmosphere. Difference between tabulated value and value for crystal obtained in Ar+O_2 is small.

Table 2. Value of spacing parameter (a), before and after irradiation

Sample	Spacing parameter a, Å	Sample	Spacing parameter a, Å
Ar (initial)	8.23453	Ar+(2%)O_2 (initial)	8.23407
Ar (irradiated)	8.23411	Ar+(2%)O_2(irradiated)	8.23405

Preliminary model of defect formation was proposed. It's supposed that gallium and oxygen vacancies are appearing in process of irradiation, further they form F- and V-centers and their complexes. Possible processes in langatate crystals under electron irradiation are shown in equations 3-8:

$$Ga^{+++} + e^- \rightarrow Ga^{++} \quad (3)$$

$$Ga^{+++} + 2e^- \rightarrow Ga^{+} \quad (4)$$

$$2Ga^{+} + O^{--} \rightarrow Ga_2O \uparrow + 2V_{Ga}^{---} \quad (5)$$

$$2O^{--} - 4e^- \rightarrow O_2 \uparrow + 2V_0^{++} \quad (6)$$

$$V_{Ga}^{---} + 3p^{+} \rightarrow V^{0} \quad (7)$$

$$V_o^{++} + 2e^- \rightarrow F^{0} \quad (8)$$

4. Conclusions

In this study significant influence of electron irradiation on optical transmission spectra of LGT crystals were shown. Dependence of transmission intensity of all absorption bands on the electron irradiation dose value was observed. Also growth atmosphere of LGT crystals influences on susceptibility to electron irradiation. In process of growth and irradiation vacancies defects are generated.

References

[1] Kaminskii A A, Mill B V, Khodzhabagyan G G 1982 Modificirovannye redkozemel'nye gallaty so strukturoy *$Ca_3Ga_2Ge_4O_{14}$* *Dokl. AN SSSR*. 264 **6** 1385 [in Russian]

[2] Shimamura K, Takeda H and Kohno T 1996 Growth and crystallization of lanthanum gallium silicate single crystals for piezoelectric application *J. Cryst. Growth* **163** 388 – 392

[3] Buzanov O A, Zabelina E V and Kozlova N S 2007 Optical Properties of Lanthanum–Gallium Tantalate at Different Growth and Post-Growth Treatment Conditions *Crystallography Reports* **52** 690-695

[4] Born M, Wolf E 1965 *Principles of Optics* (Pergamon press, 4th edition) p 720

[5] Domoroshchina E N, Kuz'michyeva G M, Rybakov V B et al. 2004 Svyaz' mejdu usloviyami vyrashchivaniya, stroeniem i opticheskimi svoistvami kristallov langasita $La_3Ga_5SiO_{14}$ *Perspectivnye materialy* **4** 17-30 [in Russian]

[6] Kaminskii A A, Mill B V, Khodzhabagyan G G et. al. 1983 Investigation of Trigonal $(La_{1-x}Nb_x)_3Ga_5SiO_{14}$ Crystals *Phys.stat. sol* (a) **80** 387-398

[7] Burkov V I, Pisarevkiy Yu V, Sizova N L, Fedorov E V and Mill B V 2007 Vliyanie γ– obluchenia na opticheskie i mechanicheskie svoystva langasita *Kristallografiya* **52** 722-725 [in Russian]

Acknowledgement

The authors gratefully acknowledge the financial support of the Ministry of Education and Science of the Russian Federation in the framework of Increase Competitiveness Program of MISiS.

IOP Conf. Series: Materials Science and Engineering **80** (2015) 012027 doi:10.1088/1757-899X/80/1/012027

Non-linear photonics for intravital microscopy in health sciences: application to detection of nanoparticles in organs

M.G. Blanchin

Institut lumière matière, UMR5306 Université Claude Bernard Lyon1-CNRS, Université de Lyon 69622 Villeurbanne cedex, France.

E-mail: Marie-Genevieve.Blanchin@univ-lyon1.fr

Abstract. Intravital microscopy (IM) for health sciences can be achieved by means of Laser Scanning Microscopy (LSM) in two Photon (2P) excitation conditions using Infrared (IR) illumination which has good penetration in all tissues and organs. Imaging is then performed using 2P emission of fluorescence (2PE) and second harmonic generation (SHG) modes which provide high contrast and good spatial resolution. These modes do reveal the different parts of organs due to differential fluorescence and SHG. In the case of atherotic aorta, for instance, the formation of the atherotic plaque can be visualised and located with respect of aorta's structure: an important application is the diagnosis about unstability or stability of the plaque. Also the spatial resolution of LSM allows revealing the heterogeneous biodistribution of iron oxide core (USPIO) nanoparticles for diagnosis of the atherotic plaque, from the aorta's scale down to the subcellular level: detection of aggregates of USPIOs at the subcellular level supports the model of USPIOs phagocytised by macrophages which target the plaque.

1. Introduction

The interest for nanoparticles (NPs) in biomedical applications is strongly growing during the last decade [1]. They are often designed as a multimodal platform for combined therapy and diagnosis (theranostics) applications [2]. Their core may contain a drug that is protected from the biological environment by a shell structure [3] like e.g. silica. The imaging moieties at the shell surface or in the core may be radioactive, paramagnetic or fluorescent for positron emission tomography (PET/SPECT) [4], magnetic resonance imaging (MRI) [5] and intravital microscopy (IM.) [6] respectively. MRI as one of the most widely used non-invasive imaging technique, permits analysis going from the whole body distribution to the microscopic biodistribution in organs and tissues, but its spatial resolution is limited at 500 μm per pixel (Figure 1). On the other hand studies of structure and biolocalisation of NPs requires techniques having spatial resolution better than a few 0.1 nm, like transmission electron microscopy (TEM). Figure1 sketches the resolution gap between MRI and TEM, for instance, which can be filled by Intravital Microscopy achieved at a resolution of a few hundred nm by means of Laser Scanning Microscopy (LSM).

2. Intravital microscopy (IM)

IM relies on the transparency of the human body in the infrared region of light spectrum (Figure 2).Major components (hemoglobin, water) of tissues and organs exhibit a minimum of absorption

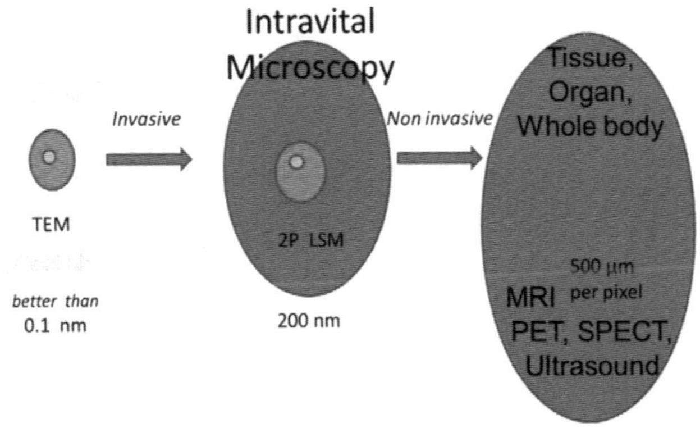

Figure 1: Two photon (2P) laser scanning microscopy (LSM) fills the resolution gap between MRI and TEM

coefficient in that region, as depicted in Figure 2. This induces a transparency window when human body parts are illuminated in the infrared range: the photon penetration can then reach several millimeters to several centimeters, depending on the organ considered.

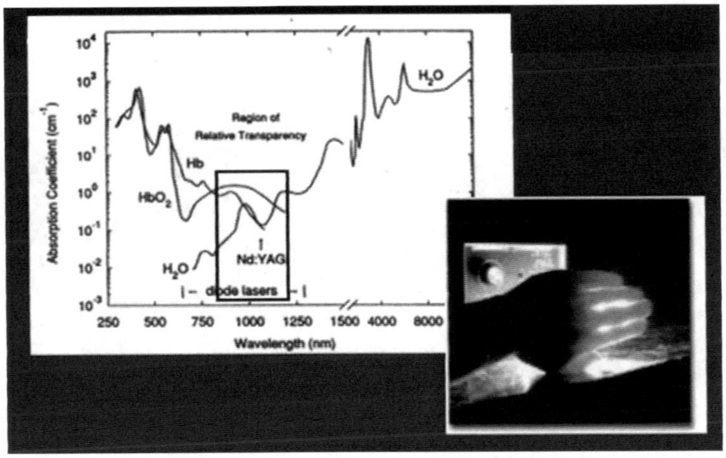

Figure 2: Variation of absorption coefficients of hemoglobin, oxyhemoglobin and water versus radiation wavelength, showing window of relative transparency in infrared (IR) range (left); example of IR radiation penetration in human hand (right)

Figure 3 illustrates an example of brain exploration achieved by IM through a skull micro window in an anesthetized mouse.

IOP Conf. Series: Materials Science and Engineering **80** (2015) 012027 doi:10.1088/1757-899X/80/1/012027

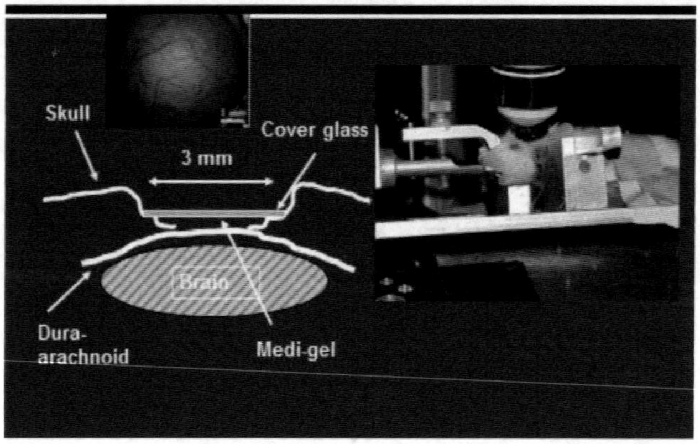

Figure 3: Exploration of mouse's brain by Intravital Microscopy

3. Non linear photonics for I.M.

Latest developments in IM thus involve non linear photonic imaging with illumination in infrared range and use of two photon fluorescence emission (2PE) and Second Harmonic Generation (SHG) modes for imaging [7,8]. Generation of these modes and corresponding energy levels are sketched on Figure 4.

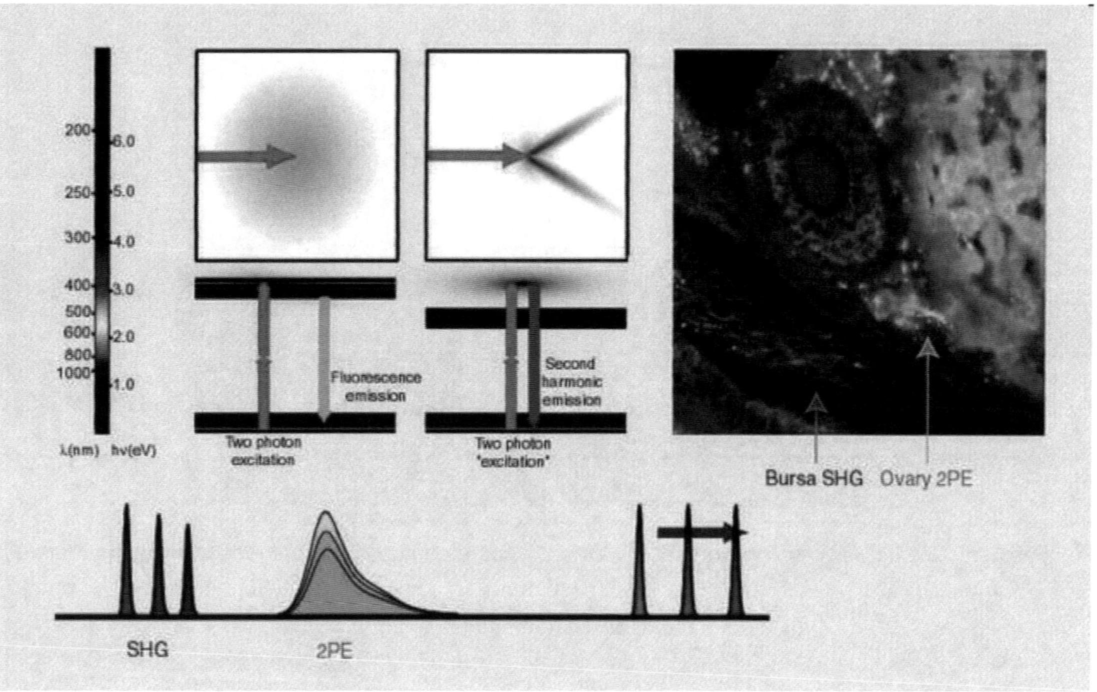

Figure 4: Two photon (2P) excitation from IR radiation giving rise to emission of fluorescence (2PE) and Second Harmonic Generation (SHG), with example of image from different organ components exciting preferentially 2PE or SHG mode.

Excitation of 2PE and SHG modes require spatial compression of photons (Figure 5) by the objective lens of a laser scanning microscope (LSM) and temporal compression during the femtosecond pulses of a femtosecond illuminating laser (Figure 5), typically a mode-locked Titanium-Sapphire laser, which produces femtosecond pulses.

The high photon density and flux required for two photon excitation typically correspond to a pulse width of approximately 100 femtoseconds and a repetition rate of about 80 MHz. The excitation wavelength can be varied between 700 and 1000 nm, a region of the specimen being illuminated by raster scanning of x-y mirrors by means of a galvanometer-driven scanner. A custom made dichroic mirror is used to maximize the reflection of the infrared radiation and to transmit the radiation in the blue-green region of the spectrum. The fluorescent radiation emitted from the sample is collected by the objective lens and transmitted to the dichroic mirror for the detection system:the most commonly used photodetectors are photomultiplier tubes due to their good sensitivity, large active areas, low price.

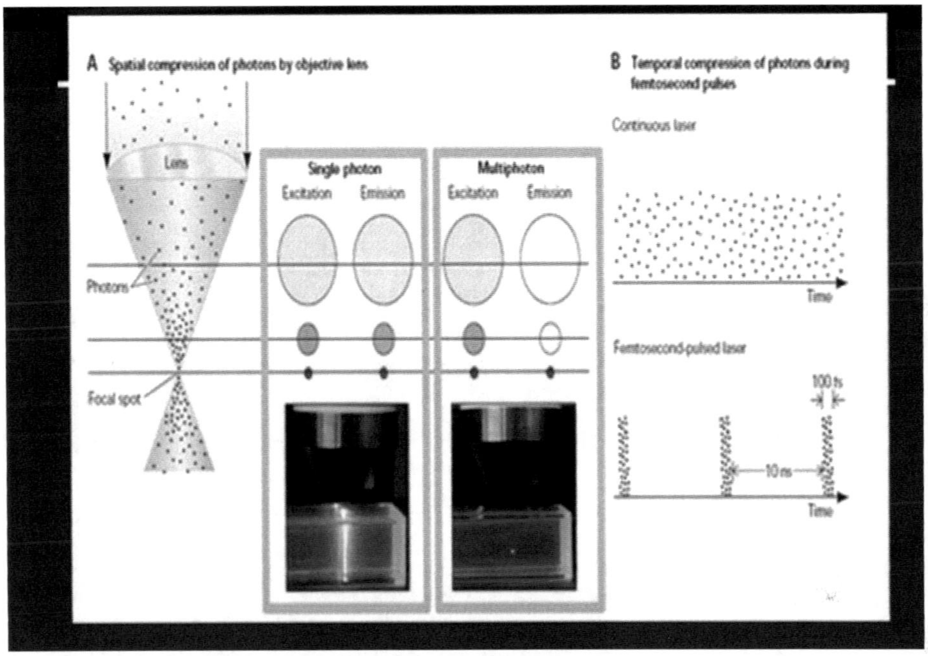

Figure 5: Temporal and spatial compression in a laser scanning microscopy (LSM) for 2P excitation

Since only ballistic photons do contribute to the image, only the specimen region corresponding to the objective focal plane is imaged and 3D imaging is obtained by 3 D reconstruction from images acquired from specimen slices successively illuminated (see section 4) [9]

4. Nanoparticles for diagnosis and /or therapy

During the past years, nanocapsules for drug targeting have been obtained from different biocompatible polymers [10 to 13]. Nanovectors (nanospheres or nanocapsules) can also be loaded with radioactive compounds for radionuclide intratumoral therapy [14] (Figure 6). Nanoparticular systems made of biopolymers encapsulating magnetic nanoparticles have been intensively studied in the last decade due to the interesting combination of their structural specifications with their magnetic properties. This combination gives rise to applications for diagnosis like Magnetic Resonance Imaging (MRI) or ultrasonic imaging, and for therapy: drug targeting, hyperthermia. In MRI studies especially, maghemite or magnetite iron oxide loaded nanosystems (Figure 6) are extensively used as contrast agents. The advantage of iron oxide nanoparticles, compared with other metallic compounds (for example gadolinium), is their relatively low toxicity [15]. The latest studies [16] show that such nanovectors can be used as specific contrast agents for targeted organs.

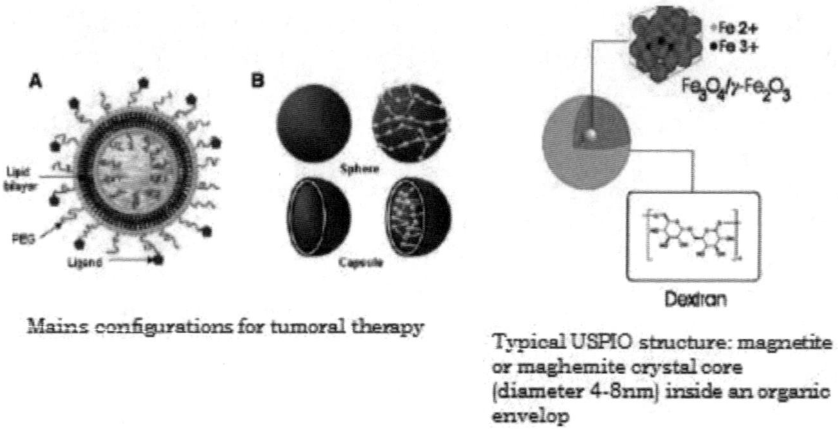

Figure 6: Structure of nanoparticles (NP) for diagnosis or/and therapy: examples

5. IM. and detection of nanoparticles in organs: validation of medical imaging

The use of ultra small superparamagnetic iron oxide (USPIO) NPs (having magnetic iron oxide core) in biomedical applications like MRI has increased significantly in the last decade, but it is not completely understood how these NP reach their target, for instance the atherotic plaque or atheroma, in diagnosis of arterosclerosis [17]. USPIOs designed as MRI contrast agents were grafted (GUERBET Labs, Paris) at the coverage surface with a dye (Rhodamine family) having emission wavelength close to 600 nm [17] and injected into atherosclerotic ApoE^{-} model mice [17] rapidly developing atherotic plaque under high-fat diet. At 32 weeks of age when the aortic plaque was well developed, the animals were administered with the same dose of 1000 µmol Fe/kg wt., as for MRI studies. Thin rings of the targeted atherotic aorta were taken from the mice sacrificed after deep anesthesia (see *Ethics* alinea herafter).

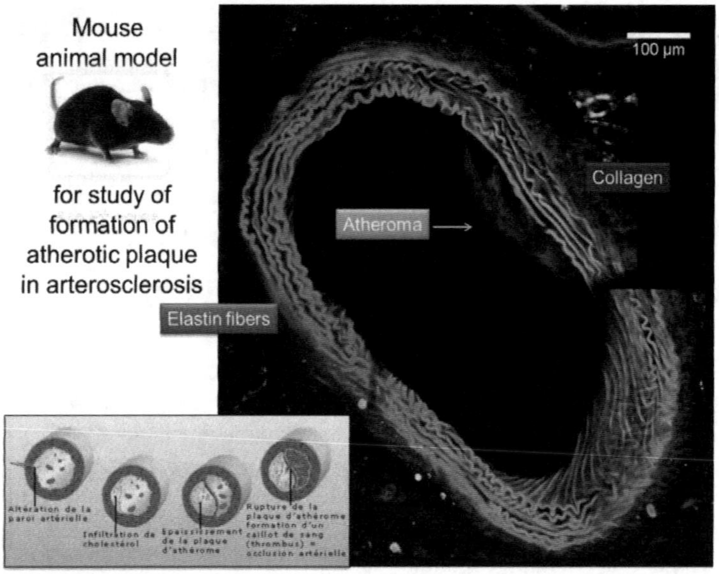

Figure 7: Overview of an atherotic aorta's ring imaged by 2PLSM, showing the aortic wall structure and the presence of atheroma (formed as sketched in left inset) in the aorta's lumen

Such aortic rings from the atherosclerotic aortas were imaged using a 2PM LSM equipment of the type described in section 3, an example of ring's overview being presented in Figure 7. The whole fluorescence emitted in the visible range 347 – 660 nm was detected, to construct such color images.

The images were acquired slice by slice in the sample depth (z-axis), depth steps of 3 to 5 μm being used to form depth stacks of several tens of slices. For a given stack, the z-projection images were obtained by averaging the pixel intensities over the depth of the stack. Orthogonal views along a given line in a given image were also recorded: they correspond to the distribution of the intensity in a slice perpendicular to the initial image plane and are displayed parallel to the chosen line with respect to the initial image. Figure 7 shows that the 2P modes of LSM does reveal the different components of the aortic wall due to fluorescence for elastin fibres and SHG blue colour for collagen fibers [17]. Also the presence of atherotic plaque is clearly visualised and located with respect of aorta's structure : an important application is the diagnosis about unstability or stability of the plaque [17]. At higher magnification the spatial resolution of LSM allows revealing the heterogeneous biodistribution of the USPIOs seen as yellow spots in the atherotic plaque (Figure 8, left). Figure 8 displays aggregates of USPIOs like in M: size and scaling of such aggregates seen at higher magnification (Figure 8, right) support the model of USPIOs phagocytised by macrophages which target the plaque.

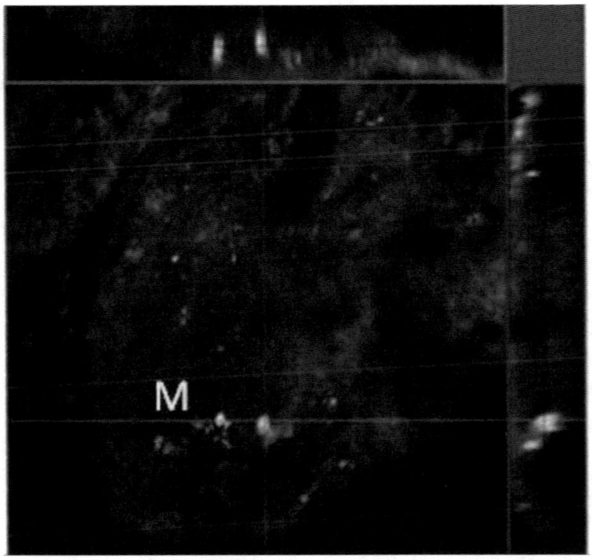

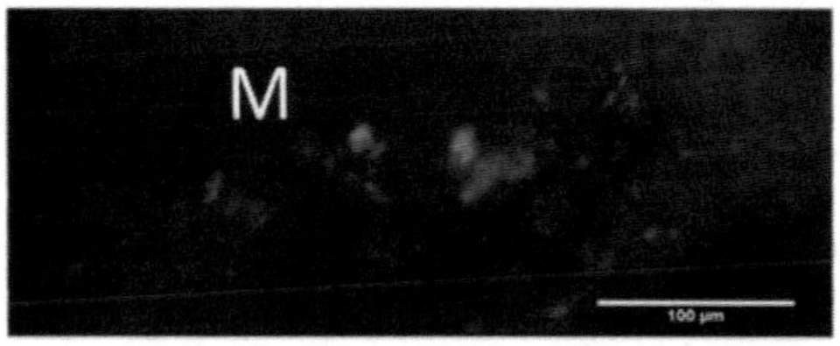

Figure 8: Biodistribution of USPIOs in atherotic plaque: plan view and orthogonal images showing USPIO's aggregates , like in M (left); high magnification of M region, consistent with USPIOs phagocytised in macrophages, bar is 100μm (right)

After such ex vivo samples were analysed by 2PLSM, specimens for TEM studies could be prepared from USPIO rich regions. Such TEM studies allowed to follow the biostructural transformations undergone by the USPIO NPs at the atomic scale [18,19]. Hence could be completed a multiscale approach from the organ's scale down to the atomic subcellular level.

6. Conclusion and perspectives

I.M. using 2PLSM permits differential imaging of organ components and tissues together with detection of NPs at the subcellular level. This allows to bridge the gap towards TEM investigations for atomic resolution. Thus 2PLSM naturally takes place in multimodal imaging platforms and supports the development of endoscopic techniques for medical imaging.

Ethics: all the animal experimentations mentioned in the present paper were conducted according to European and French ethical regulations

Acknowledgements: Special thanks are due to Dr B. van der Sanden, Platform Intravital Microscopy, France Life Imaging, CLINATEC, Grenoble (France). GUERBET Laboratories (Paris, France) are gratefully acknowledged for providing USPIO nanoparticles.

References

[1] A.M. Morawski, G.A. Lanza, S.A. Wickline, *Current opinion in biotechnology* **16** (2005) 89
[2] D.B. Shieh, F.Y. Cheng, C.H. Su, C.S. Yeh, M.T. Wu, Y.N. Wu, C.Y. Tsai, C.L. Wu, D.H. Chen, C.H. Chou, *Biomaterials* **26** (2005) 7183
[3] M. Timko, M. Koneracka, N. Tomasovicova, P. Kopcansky, V. Zavisova, *J. Magn. Mater.* **300** (2006) 191
[4] J.M. Ollinger, J.A.Fessler, *IEEE Signal Processing Magazine*, January (1997) 1
[5] A.K. Gupta, M. Gupta, *Biomaterials* **26** (2005) 3995
[6] W. Yu, J.C. Braz, A.M. Dutton, P.Prusakov, M. Rekhter, *J Biomed Opt.*, **12** (2007) 054008
[7] P.T.C. So, C.Y. Dong, B.R. Masters, K. M. Berland., *Ann.. Rev. Biomed. Eng.*, **2** (2000) 399
[8] W. Denk, J.H.Strickler, W.W. Webb, *Science,* **248** (1990) 73
[9] M. Maurin, O. Stephan, J.C. Vial, S.R. Marder, B. Van der Sanden, *J Biomed Opt.*, **16** (2012) 036001
[10] W. Zheng, F. Gao, H. Gu, *J. Magn. Magn. Mater.* **288** (2005) 403
[11] F.X. Hu, K.G. Neoh, E.T. Kang, *Biomaterials,* **27** (2006) 5725
[12] S.J. Lee, J.R. Jeongb, S.C. Shinb, J.-C. Kimc, Y.-H. Changd, K.-H. Leea, J.-D. Kim, *Colloids Surf. A* **255** (2005) 19
[13] M. Hamoudeh, H. Fessi, *J. Colloid Interface Sci.* **300** (2006) 584
[14] M. Hamoudeh, H. Fessi, *European Journal of Pharmaceutics and Biopharmaceutics*, **67** (2007) 597
[15] A. Iannone, R.L. Magin, T. Walczack, M. Federico, H.M. Swartz, A. Tomasi, V. Vannini, *Magn. Reson. Med.* **22** (1991) 435
[16] C.M. Lee, H.J. Jeong, S.L. Kim, E.M. Kim, D.W. Kim, S.T. Lim, K.Y. Jang, Y.Y. Jeong, J.W. Nah, M. Sohn, *Int J Pharm* **371** (2009) 163
[17] V.A. Maraloiu, M.G. Blanchin, *Multiscale study of magnetic nanovectors* (2013) Scholar's Press
[18] J.D. Lopez-Castro, V.A. Maraloiu, J.J. Delgado, J.J. Calvino, M.G. Blanchin , J.M. Dominguez-Vera,., *Nanoscale*, **3** (2011) 4597
[19] V.A. Maraloiu, F. Appaix, A. Broisat, D. Leguellec, V.S. Teodorescu, C.Ghezzi, B. Van der Sanden, M.G. Blanchin (2015), submitted

AUTHOR INDEX

Institute of Physics
Dirac House, Temple Back
Bristol BS1 6BE UK

CURRAN ASSOCIATES INC
proceedings.com